Selected Vitamins, Minerals, and Functional Consequences of Maternal Malnutrition

Volume Editor

Artemis P. Simopoulos

The Center for Genetics, Nutrition and Health,
American Association for World Health, Washington, D.C.

35 figures and 24 tables, 1991

KARGER

Basel · München · Paris · London · New York · New Delhi · Bangkok · Singapore · Tokyo · Sydney

World Review of Nutrition and Dietetics

Library of Congress Cataloging-in-Publication Data
    Selected vitamins, minerals, and functional consequences of maternal malnutrition /
    volume editor, Artemis P. Simopoulos.
    (World review of nutrition and dietetics; vol. 64)
    Includes bibliographical references.
    Includes index.
    1. Vitamins in human nutrition.   2. Minerals in human nutrition.   3. Infants – Nutrition.
    4. Malnutrition in pregnancy.
    I. Simopoulos, Artemis P., 1933–   .   II. Series.
    [DNLM: 1. Ascorbic Acid Deficiency.   2. Calcium – metabolism.
    3. Infant, Newborn Diseases – etiology.   4. Nutrition Disorders – complications.
    5. Nutrition Disorders – in pregnancy.   6. Vitamin A Deficiency   7. Vitamin K Deficiency.]
    ISBN 3–8055–5168–1

Bibliographic Indices
    This publication is listed in bibliographic services, including Current Contents® and Index
    Medicus.

Drug Dosage
    The authors and the publisher have exerted every effort to ensure that drug selection and dosage
    set forth in this text are in accord with current recommendations and practice at the time of
    publication. However, in view of ongoing research, changes in government regulations, and the
    constant flow of information relating to drug therapy and drug reactions, the reader is urged to
    check the package insert for each drug for any change in indications and dosage and for added
    warnings and precautions. This is particularly important when the recommended agent is a new
    and/or infrequently employed drug.

©   Copyright 1991 by S. Karger AG, P.O. Box, CH–4009 Basel (Switzerland)
    Printed in Switzerland by Thür AG Offsetdruck, Pratteln
    ISBN 3–8055–5168–1

Selected Vitamins, Minerals,
and Functional Consequences of Maternal Malnutrition

World Review of Nutrition and Dietetics

Vol. 64

Series Editor
Artemis P. Simopoulos, Washington, D.C.

KARGER

Basel · München · Paris · London · New York · New Delhi · Bangkok · Singapore · Tokyo · Sydney

Contents

Retinoids in the Host Defense System

Vitamin K Deficiency in Infancy

Comparative Properties of Erythrocyte Calcium-Transporting Enzyme in Different Mammalian Species

Enitan A. Bababunmi, Olufunso O. Olorunsogo, Ibadan, *Clement O. Bewaji,* Ilorin 109

Functional Consequences of Maternal Malnutrition

Teresa González-Cossío, Hernán Delgado, Guatemala City 139

Contents

Preface

Extensive research on vitamins and minerals over the past 10 years indicates that certain vitamins behave like hormones in their function (vitamin A and vitamin D) and that the original studies that led to the discovery of vitamins based on the deficiency disease model were only the beginnings of the recognition of their importance to health.

The 90s are ushering in the Golden Epoch of Nutrition because of advances in molecular biology and the new studies on the role of nutrients in gene expression. Vitamin A is not only a substance which prevents xerophthalmia and ascorbic acid does not prevent scurvy only. In fact, it could be said that the discoveries that vitamins have antioxidant properties and that certain life-styles increase the vitamin requirements (i.e. smoking increases the requirement of vitamin C because of interference with its absorption by the gastrointestinal tract of the smoker, and the increased catabolic rate) are leading the way for re-examining our approach in determining the nutritional requirements of vitamins. We need to consider the effects of vitamin C, vitamin A and carotenoids in growth, cell differentiation, and in tumor (cancer) development. We know that the human retinoic acid receptor belongs to the family of nuclear receptors which are similar to thyroid and steroid hormones and that retinoids can control genomic expression of cells.

We now begin to re-examine our position both on the role of certain vitamins in cancer prevention, in aging, and in determining the appropriate dosage under various conditions, age and life-style changes. The first two papers, 'Vitamin C as an Antioxidant' by Etsuo Niki and 'Smoking and Vitamin C' by Akira Murata, provide extensive review of the literature and new data on vitamin C. The third paper on 'Retinoids in the Host Defense System' by Manabu Yamamoto reviews the role of retinoids in disease prevention, morphological changes in lymphoid organs with vitamin A deficiency, immune competence and retinoids, and mechanisms of action.

Hemorrhagic disease of the newborn is a preventable condition that responds to vitamin K administration. Hemorrhagic disease is no longer considered a single disorder. Three specific types or groups [early hemorrhagic disease of the newborn (HDN), classical HDN, and late neonatal hemorrhagic disease (late neonatal vitamin K deficiency)], based on the time (age) of occurrence are reviewed in the fourth paper on 'Vitamin K Deficiency in Infancy' by Ichiro Matsuda et al. Reference is also made to a genetic variant (an inborn deficiency of vitamin K epoxide reductase), and the fact that the incidence of the third type is much more prevalent in Japan than in Europe and the US. Does it have a genetic basis or is it another example of gene-nutrient interaction being more prevalent in one ethnic group?

Calcium is the most common cation in the body. Furthermore calcium, hormones, and cyclic nucleotides are the most important regulators or messengers in mammalian systems and their activities are interwoven. In the paper 'Comparative Properties of Erythrocyte Calcium-Transporting Enzyme in Different Mammalian Species' the comparative properties of erythrocyte calcium-transporting enzyme in different mammalian species are presented and discussed critically by Enitan Bababunmi et al., particularly the Ca^{2+}-transporting ATPases. The authors believe that characterization of these proteins is particularly important to Africans in the understanding of the pathology of malnutrition, parasitic diseases, genetic disorders and cardiovascular disease, hypertension and sickle cell anemia.

Malnutrition in pregnancy adversely affects fetal development as well as lactation. Alterations in hemodynamic processes, such as low blood pressure, have been observed in undernourished women. Questions remain if hormonal production is altered in undernourished women and whether hormonal changes influence the changes in hemodynamic processes. In the paper 'Functional Consequences of Maternal Malnutrition' by Teresa González-Cossío and Hernán Delgado, the relationship of calcium, protein, maternal weight, blood pressure, fetal weight, hormonal and placental changes, and milk production are extensively reviewed and critiqued. Policy makers should be particularly interested in this paper as well as researchers in maternal nutrition and in the growth and development of the fetus and infant.

This volume should be of interest to nutrition researchers, molecular biologists, physiologists, physicians (particularly pediatricians and obstetricians), general nutritionists, dietitians and policy makers.

Artemis P. Simopoulos, MD

Simopoulos AP (ed): Selected Vitamins, Minerals, and Functional Consequences of Maternal Malnutrition. World Rev Nutr Diet. Basel, Karger, 1991, vol 64, pp 1–30

Vitamin C as an Antioxidant

Etsuo Niki

Department of Reaction Chemistry, Faculty of Engineering,
The University of Tokyo, Hongo, Tokyo, Japan

Contents

Vitamin C, ascorbic acid, has versatile functions in biological systems such as the facilitation of collagen formation and iron absorption [1–6]. Vitamin C is a water-soluble vitamin and it is an essential nutrient in man, monkey and guinea pig, which do not have the ability to synthesize the compound. It has been known for some time that ascorbic acid also functions as an antioxidant and protects against oxidative deterioration of vegetable oils, animal fats, citrus oils, and fat-containing foods such as fish, margarine, and milk [7, 8]. These functions of vitamin C are derived largely from its redox properties. Numerous reports have been published by many investigators on the role of vitamin C as an antioxidant, but the detailed mechanisms for the antioxidant actions by ascorbic acid are not clearly understood and frequently controversial.

Ascorbic acid, AH_2 (fig. 1, *1*), is a weak, dibasic acid and its pKa is 4.2, suggesting that at pH 7 or under physiological conditions it is pres-

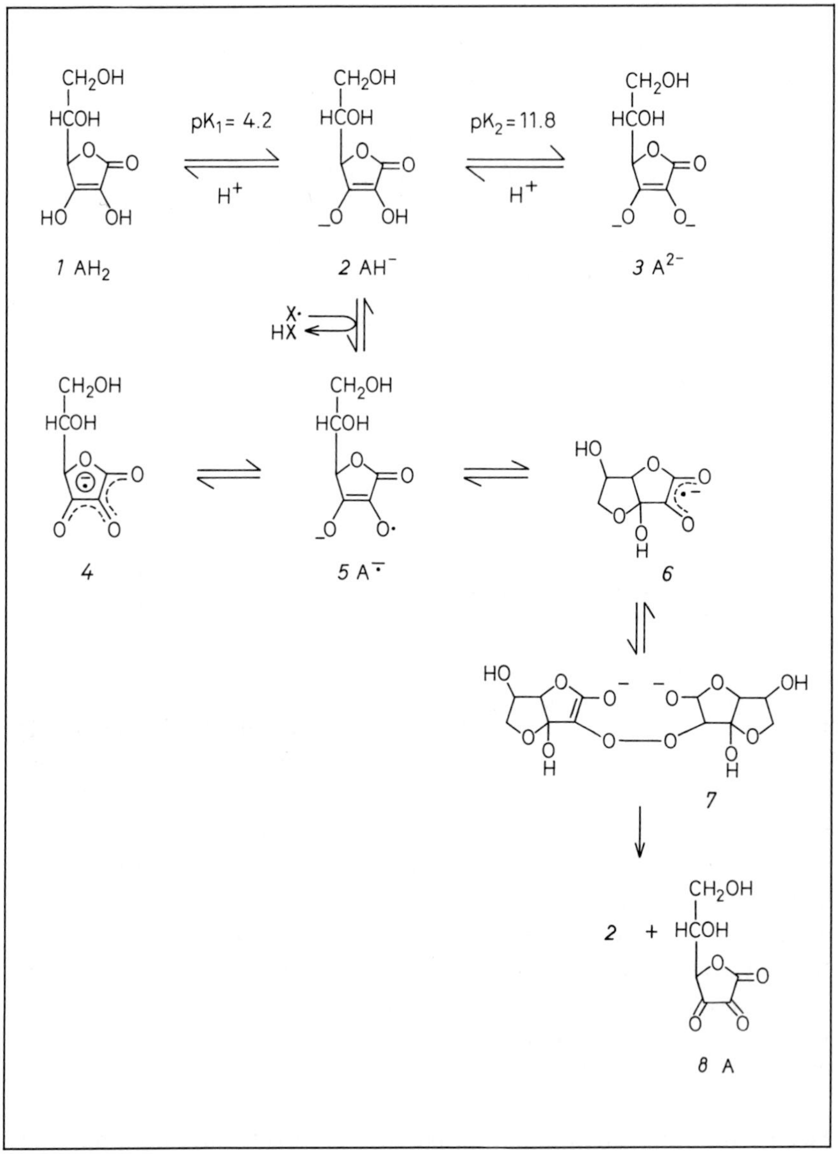

Fig. 1. Oxidation of ascorbic acid.

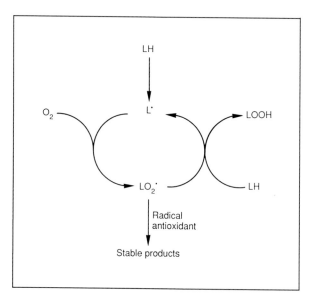

Fig. 2. Free radical-mediated chain oxidation of lipids.

ent predominantly as an ascorbate anion, AH⁻ (fig. 1, 2) (reaction 1). The NMR study demonstrates that deprotonation occurs at the C_3 hydroxyl to give the AH⁻ anion [9]. Ascorbate can undergo a two-step reversible oxidation process (reactions 2, 3) to form dehydroascorbic acid, A (fig. 1, 8), with the formation of ascorbyl radical A⁻ (fig. 1, 5) as an intermediate.

$$AH_2 \longrightarrow AH^- + H^+ \tag{1}$$

$$AH^- \longrightarrow A^{\overline{\cdot}} + H^+ \tag{2}$$

$$A^{\overline{\cdot}} \longrightarrow A \tag{3}$$

The formation of ascorbyl radical may proceed via a hydrogen atom abstraction or by electron transfer followed by rapid deprotonation. The pK of ascorbyl radical is –0.45 and it is present in its anionic form in the pH range of 0–13. The unpaired electron in the ascorbyl radical is delocalized over a highly conjugated tricarbonyl system (fig. 1, 4), which makes ascorbyl radical unreactive. It decays either by disproportionation or by reaction with other radicals. The ESR spectrum of the ascorbyl radical is well characterized. Dehydroascorbic acid has been regarded as a tricarbonyl compound with a C_1–C_4 lactone ring, but now it is accepted that it may exist in a variety of forms with the hydrated hemiketal being favored in aqueous solution.

Autoxidation of Fats, Oils, and Biological Molecules

Foods containing polyunsaturated fatty acids and their esters are readily oxidized by molecular oxygen in the air and deteriorated: smells develop and taste turns stale. Such an oxidation, called autoxidation, proceeds by a free radical chain mechanism as illustrated in figure 2 [10]. In the initiation step, a lipid radical, L·, is formed from lipid, LH, usually by an influence of light, heat, metal, and irradiation. There are also external sources of free radicals such as environmental pollutants, cigarette smoke and car exhaust fumes. The carbon-centered, lipid radical reacts with oxygen rapidly to give lipid peroxyl radical, $LO_2^·$, which attacks another lipid molecule and abstracts a hydrogen atom to give lipid hydroperoxide and, at the same time, a new lipid radical that starts the propagation sequence over again. Thus, many molecules of lipids may be oxidized to lipid hydroperoxides for every initiation event, and this oxidation is called free radical chain oxidation. The propagation cycle is broken by termination reactions. The free radicals are destroyed by an interaction with other radicals or by interacting with chain-breaking antioxidants, which scavenge radicals and interrupt chain reactions.

The hydroperoxide formed as a primary product of lipid oxidation may serve as a radical source, hence the autoxidation is usually autocatalytic as shown in figure 3.

The free radical-mediated oxidations of biological molecules, membranes and tissues in vivo have received much attention recently and it is now generally accepted that they are closely connected with a variety of pathological events, cancer, and even aging processes [11–14]. Thus, the autoxidations of foods and biological molecules are usually deleterious processes and should be prevented. The aerobic organisms are protected from free radical-mediated oxidations by an array of defence systems [15, 16], thanks to which we do not get rancid like foods. This implies that much can be learned from biological systems and applied for the prevention of oxidative deterioration of foods.

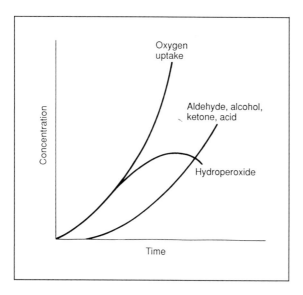

Fig. 3. The time course of the oxidation of lipids.

Role of Vitamin C and its Derivatives as Antioxidants

Ascorbic acid functions as an antioxidant by stabilizing active oxygen species which induce and carry the chain oxidations of foods and biological molecules. Probably the most important reaction in the inhibition of oxidation by ascorbic acid must be the scavenging of oxygen radicals such as hydroxyl, hydroperoxyl, lipid peroxyl, and lipid alkoxyl radicals.

It is reported that ascorbic acid reacts with hydroxyl radical at a rate constant of $7.2 \times 10^9 - 1.3 \times 10^{10}$ M^{-1} s^{-1} [17, 18] which shows that the reaction is very fast and diffusion-controlled. However, this does not mean that ascorbic acid is a specific hydroxyl radical scavenger, since hydroxyl radical is so reactive that it can react with many other compounds at about the same rate. The reaction proceeds either by an electron transfer or by the addition of hydroxyl radical to the double bond of ascorbic acid [19]. The rate constants for the reactions of ascorbic acid with some oxygen radicals are shown in table 1 [20].

Ascorbic acid reacts with superoxide at a rate constant of 10^4–10^5 M^{-1} s^{-1} [21]. In aprotic media, ascorbic acid is oxidized to dehydroascorbic acid by superoxide with a second-order rate constant of 2.8×10^4

Table 1. Absolute rate constants for the reaction of radicals with ascorbic acid (pH 7.0) [20]

Radicals	k, M^{-1} s^{-1}
$Cl_3COO^{\bullet}$	2.0×10^8
$Cl_2CHOO^{\bullet}$	2.6×10^8
$ClCH_2OO^{\bullet}$	9.2×10^7
$^{\bullet}OOCCl_2CO_2^-$	9.0×10^7
$^{\bullet}OOCHClCO_2^-$	5.1×10^7
$CH_3OO^{\bullet}$	2.2×10^6
$(CH_3)_2C(OH)CH_2OO^{\bullet}$	2.1×10^6

M^{-1} s^{-1}. Dehydroascorbic acid is oxidized by superoxide to give oxalate ion and the anion of threonic acid [22]. Ascorbic acid also scavenges hydroperoxyl radical at a rate constant of 1.6×10^4 M^{-1} s^{-1} [17, 18].

$$O_2^{\bar{\cdot}} + AH^- \longrightarrow HO_2^- + A^{\bar{\cdot}} \tag{4}$$

$$HO_2^{\bullet} + AH^- \longrightarrow H_2O_2 + A^{\bar{\cdot}} \tag{5}$$

Singlet oxygen reacts with ascorbic acid at a rate constant of 8.30×10^6 M^{-1} s^{-1} [23].

Polyunsaturated fatty acids, their esters, and phosphatidylcholines from foods such as egg yolk and soybean are stable at ambient temperatures in the dark, and their oxidations proceed only very slowly. However, if the radicals are formed, they readily attack lipids and induce a free radical chain oxidation. Free radicals can be formed by various reactions such as photochemical, thermal, and metal-catalyzed decompositions of hydroperoxides. A radical initiator such as 2,2'-azobis(2,4-dimethylvaleronitrile) (AMVN) or 2,2'-azobis(2-amidinopropane)dihydrochloride (AAPH) generates free radicals at a known rate and site. Therefore, azo initiators such as AMVN and AAPH are often used in the laboratory for fundamental and kinetic studies [24, 25]. For example, methyl linoleate is oxidized in the presence of an azo initiator at 37 °C and constant rates of oxygen uptake and conjugate diene hydroperoxide formation are observed [10, 26, 27]. The oxidations proceed by a sequence of reactions 6 to 11, where A–N=N–A is an azo initiator, LH is a lipid, and $L^{\bullet}$ and $LO_2^{\bullet}$ are lipid radical and lipid peroxyl radical, respectively. When a chain-breaking antioxidant such as ascorbic acid is added to the solution, it scavenges $AO_2^{\bullet}$ and/or $LO_2^{\bullet}$ radicals and suppresses the oxidation.

Chain initiation:

$$A-N=N-A \longrightarrow A^\cdot + N_2 + {}^\cdot A \tag{6}$$

$$A^\cdot + O_2 \longrightarrow AO_2^\cdot \tag{7}$$

$$AO_2^\cdot + LH \xrightarrow{O_2} AOOH + LO_2^\cdot \tag{8}$$

Chain propagation:

$$LO_2^\cdot + LH \longrightarrow LOOH + L^\cdot \tag{9}$$

$$L^\cdot + O_2 \longrightarrow LO_2^\cdot \tag{10}$$

Chain termination:

$$2\,LO_2^\cdot \longrightarrow stable\ products \tag{11}$$

Figure 4 shows the effect of ascorbic acid on the oxidation of methyl linoleate in tert-butyl alcohol/methanol (3:1 by volume) initiated with 10 mM AMVN [28]. In the absence of ascorbic acid, the oxidation proceeds at a constant rate without any noticeable induction period. The addition of ascorbic acid suppresses the oxidation and produces a clear induction period. The higher the ascorbic acid concentration, the longer the induction period and the smaller the rate of oxidation during the induction period. As shown in figure 5, the length of induction period is directly proportional to the concentration of ascorbic acid.

Under these conditions, ascorbic acid scavenges peroxyl radicals and interrupts the chain propagation and the chain termination proceeds by the reaction 12 instead of reaction 11,

$$n\,LO_2^\cdot + IH \xrightarrow{k_{inh}} stable\ products \tag{12}$$

where IH is ascorbic acid and n is a stoichiometric number of peroxyl radicals scavenged by each antioxidant. The induction period, t_{inh}, is given by equation 13, where R_i is the rate of chain initiation. The equation 13 suggests that the induction period is proportional to the antioxidant concentration, in agreement with the results shown in figure 5.

$$t_{inh} = n[IH]/R_i \tag{13}$$

$$R_{inh} = \frac{k_p[LH]R_i}{nk_{inh}[IH]} \tag{14}$$

$$\frac{k_{inh}}{k_p} = \frac{[LH]}{R_{inh}t_{inh}} \tag{15}$$

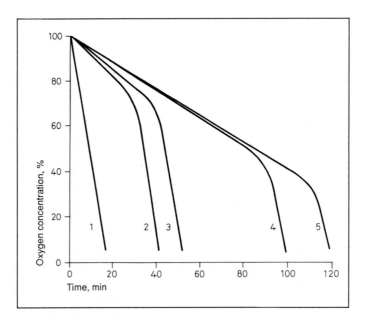

Fig. 4. Inhibition of oxidation of methyl linoleate in tert-butyl alcohol/methanol (3:1 by volume) by vitamin C at 37 °C, [AMVN] = 10 mM [28]

	Curve				
	1	2	3	4	5
[Vitamin C], µM	0	41	55	110	166
t_{inh}, s	0	1860	2430	5460	6750

The rate of oxidation during the induction period, R_{inh}, is given by the equation 14. The equations 13 and 14 give equation 15, from which the ratio of the rate constants, k_{inh}/k_p, can be calculated without knowing n value [28, 29]. This ratio is important since it determines the efficacy and activity of antioxidant. The data in figure 4 give $k_{inh}/k_p = 7.5 \times 10^2$ for ascorbic acid. The absolute rate constant k_p for methyl linoleate was determined by Howard and Ingold [26] as $k_p = 62\ M^{-1}\ s^{-1}$ at 30 °C and it was estimated as $k_p = 230\ M^{-1}\ s^{-1}$ at 50 °C [27]. The interpolation of these values gives $k_p = 100\ M^{-1}\ s^{-1}$ at 37 °C, from which k_{inh} is calculated as $7.5 \times 10^4\ M^{-1}\ s^{-1}$.

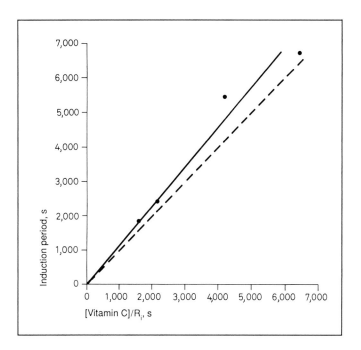

Fig. 5. Plot of induction period against [vitamin C]/R_i in the oxidation of methyl linoleate in tert-butyl alcohol/methanol (3:1 by volume) at 37 °C inhibited by vitamin C. The dotted line corresponds to slope = 1.0 [28].

The value of n is important since it is closely connected with the mechanism of inhibition of oxidation by the antioxidant. Wayner et al. [30] found recently that the n value for ascorbic acid depended on the concentration of ascorbic acid. It was indicated [30] that n approached to 2.0 as ascorbic acid concentration was decreased to 0 and that n approached to 0 as ascorbic acid concentration was increased. This variation has been attributed to the fact that ascorbate not only acts as a chain-breaking antioxidant but can also undergo autoxidation by equations 16 and 17 [30]. The rate constants for the reactions 16 and 17 are reported as 600 M^{-1} s^{-1} and 10^5 M^{-1} s^{-1} respectively [19]. The autoxidation of ascorbate is slowed in the presence of superoxide dismutase, SOD [31].

$$A^{\cdot -} + O_2 \longrightarrow A + O_2^{\cdot -} \tag{16}$$

$$O_2^{\cdot -} + AH^- \longrightarrow HOO^- + A^{\cdot -} \tag{17}$$

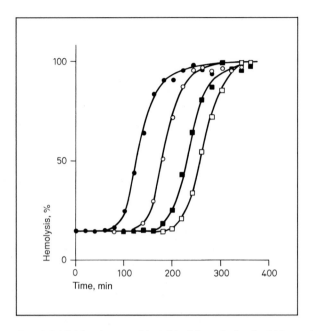

Fig. 6. Inhibition by ascorbic acid of hemolysis of rabbit erythrocytes induced by 70 mM AAPH at 37 °C. [Ascorbic acid]: ● = 0; ○ = 100; ■ = 200; □ = 300 μM [15].

Ascorbic acid also suppresses the oxidations of methyl linoleate micelles and soybean phosphatidylcholine (PC) liposomes in the aqueous dispersions initiated with AAPH which generates free radicals in the aqueous region [24, 32, 33].

The free radicals generated from AAPH in the aqueous suspensions of erythrocytes attack the erythrocyte membranes, induce the oxidations of membrane lipids and proteins, and eventually cause hemolysis [15, 34, 35]. The hemolysis takes place sooner with increasing AAPH concentration [15, 35]. As shown in figure 6, ascorbic acid present in the same aqueous suspensions suppresses the radical-induced hemolysis dose-dependently [15]. Apparently, ascorbic acid scavenges oxygen radicals formed from AAPH in the aqueous phase before the radicals attack erythrocyte membranes and thus it protects membranes from oxidative damage.

The blood plasma has a very large concentration of radical trapping antioxidants [36, 37]. Ingold and his colleagues [36, 37] have measured the

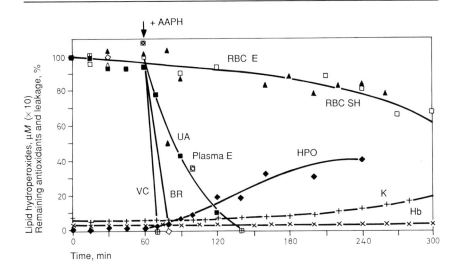

Fig. 7. Decrease of ascorbic acid (VC, ⊞), bilirubin (BR, ◙), uric acid (UA, △), plasma α-tocopherol (Plasma E, ■), membrane α-tocopherol (RBC E, □), and membrane sulfhydryl groups (RBC SH, ▲), formation of neutral lipid hydroperoxides (HPO, ♦), and leakages of hemoglobin (Hb, ×) and potassium ion (K, +) in the oxidation of 4 times diluted human whole blood with physiological saline initiated with 30 mM AAPH at 37 °C under air [38].

total radical-trapping antioxidant potential (TRAP) in the human plasma and found that vitamin C, vitamin E, urate and protein act as the major components.

It has been found recently that ascorbic acid functions as a primary defence in the whole blood against free radicals [38]. When the radicals are generated from AAPH in the aqueous phase in the whole blood suspensions, the radicals attack lipids and proteins both in the plasma and in the erythrocyte membranes. Various antioxidants in the plasma and erythrocyte scavenge the radicals and suppress the oxidative damage. As shown in figure 7, ascorbic acid scavenges the radicals most efficiently and it is consumed most rapidly. After ascorbic acid is depleted, bilirubin, uric acid, vitamin E in the plasma are consumed. Vitamin E and the S-H groups in the erythrocyte membranes are consumed after most antioxidants in the plasma are consumed. These results suggest that vitamin C functions as a

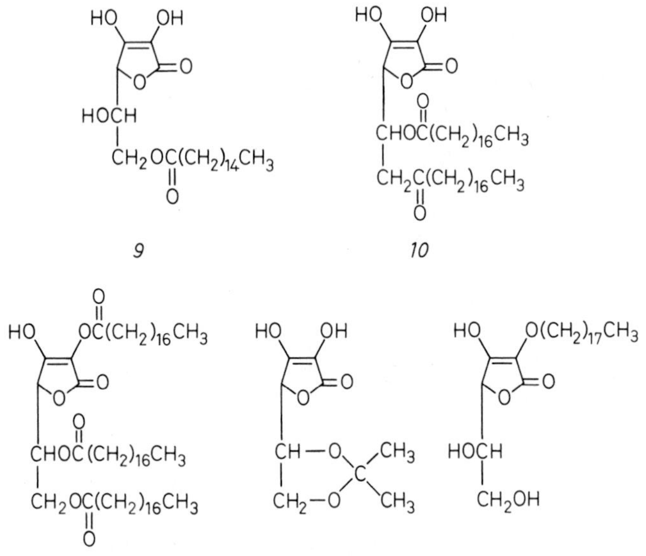

Fig. 8. Lipid-soluble derivatives of ascorbic acid.

primary and the most important defence against the radicals in the aqueous phase.

The action of vitamin C in vivo has been also observed [39, 40]. For example, Kunert and Tappel [39] found that the exhalation of pentane and ethane, a measure of the in vivo lipid peroxidation, increased markedly when carbon tetrachloride was introduced into vitamin C deficient rat. Kato et al. [40] found that ascorbic acid suppressed the lipid peroxidation of rat liver induced by polychlorinated biphenyls.

It must be noteworthy that, although ascorbic acid functions as a potent, chain-breaking antioxidant in the aqueous phase, it can not scavenge the radicals within the lipid region of the membranes efficiently [32, 33]. Thus, ascorbic acid can suppress the oxidations of membranes when they are induced by water-soluble AAPH and the radicals are formed ini-

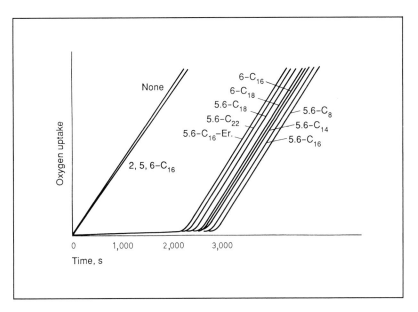

Fig. 9. Inhibition of oxidation of 145 mM methyl linoleate in solution by fatty acid esters of ascorbic acid at 37 °C under air initiated with AMVN [41].

tially in the aqueous phase, but it can not suppress the oxidations of membranes when they were induced by the radicals generated from hydrophobic AMVN.

Ascorbic Acid Esters

Several kinds of lipid-soluble derivatives of ascorbic acid have been prepared and their antioxidant activities have been studied [7, 41]. Some of them (9–13) are shown in figure 8.

Cort [7] found that ascorbic acid esters such as 6-ascorbyl palmitate (fig. 8, 9) functioned as antioxidants for vegetable oils, animal fats, and other fat-containing foods. Similar results have been reported elsewhere [42]. Various kinds of fatty acid esters of L-ascorbic acid has been synthesized recently and their role as antioxidants has been studied [41]. As shown in figure 9, the fatty acid esters of ascorbic acid at 6- or at both 5- and 6-positions were effective as antioxidant and produced clear induction

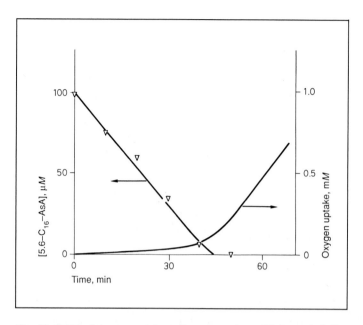

Fig. 10. Rates of oxygen uptake and consumption of 5,6-ascorbyl distearate in the oxidation of 159 m*M* methyl linoleate in benzene initiated with 10.1 m*M* at 37 °C under air [41].

period, whereas those having ester group at 2-position did not act as an antioxidant, suggesting that the O–H group at the 2-position plays an important role. 5,6-Erythorbyl dipalmitate was also effective in suppressing the oxidation. Figure 10 shows that the ascorbic acid ester was consumed linearly with time during the induction period and when it was completely depleted, the induction period was over and the rate of oxidation increased to the same level as that without ascorbic acid ester [41].

Synergistic Inhibition of Oxidation by Vitamin C and Vitamin E

One of the important features of ascorbic acid is that it can act as a synergist in the inhibition of oxidation [8, 15, 43–46]. Especially, the synergism between ascorbic acid and α-tocopherol (vitamin E) in the inhibition of lipid peroxidation has been known [28, 32, 33, 41, 47–58] and now well established [8, 15, 43–46].

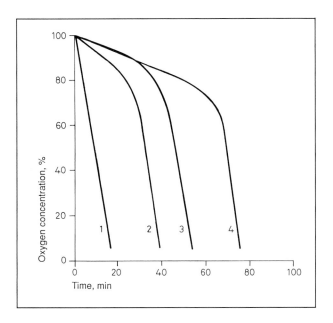

Fig. 11. Inhibition of oxidation of methyl linoleate in tert-butyl alcohol/methanol (3:1 by volume) by vitamin E and vitamin C, [AMVN] = 0.010 M [28].

Figure 11 shows the results of oxidation of methyl linoleate in the presence and absence of α-tocopherol and ascorbic acid [28]. In the absence of an antioxidant, the oxidation proceeds smoothly without any induction period (curve 1). The addition of either ascorbic acid (curve 2) or α-tocopherol (curve 3) suppresses the oxidation markedly and produces a clear induction period. The smaller rate of inhibited oxidation in the presence of α-tocopherol is attributed to a higher inhibition rate constant k_{inh} and higher n of α-tocopherol than ascorbic acid. When both ascorbic acid and α-tocopherol are added (curve 4), the induction period is lengthened to the sum of the induction period observed when either α-tocopherol or ascorbic acid is used alone. What is interesting is that the rate of oxidation is the same throughout the induction period as the rate inhibited by α-tocopherol alone. In fact, the rate constant $k_{inh} = 4.0 \times 10^5$ M^{-1} s^{-1} obtained in this system is much larger than $k_{inh} = 7.5 \times 10^4$ M^{-1} s^{-1} for ascorbic acid and close to that observed for α-tocopherol, $k_{inh} = 5.1 \times 10^5$ M^{-1} s^{-1}.

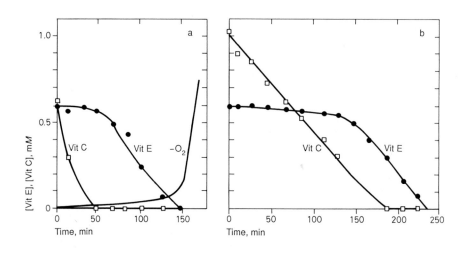

Fig. 12. Disappearance of vitamin E (●) and vitamin C (□) in the oxidation of methyl linoleate at 37 °C in tert-butyl alcohol/methanol (3:1 by volume). [LH] = 0.60 *M*, [AMVN] = 0.010 *M*, [vitamin E] = 0.595 m*M*, [vitamin C] = 0.620 m*M* (a) and 1.03 m*M* (b) [28].

The rate of consumption of α-tocopherol and ascorbic acid during the oxidation of methyl linoleate is also interesting. When either ascorbic acid or α-tocopherol is used alone, they are consumed linearly with time. However, as shown in figure 12, when both α-tocopherol and ascorbic acid are used, only ascorbic acid disappears at first and α-tocopherol remains almost constant, and α-tocopherol begins to decrease after ascorbic acid is depleted [28].

Figure 13 shows that similar results are observed for α-tocopherol and fatty acid ester of ascorbic acid in the oxidation of methyl linoleate in chloroform solution. 5,6-Ascorbyl distearate disappeared first linearly with time whereas α-tocopherol remained almost constant and it began to decrease after ascorbic acid ester was depleted [41].

These results suggest that α-tocopherol which has higher k_{inh} than ascorbic acid scavenges the peroxyl radical more quickly than ascorbic acid, but the α-tocopheroxyl radical derived from α-tocopherol is reduced back to regenerate α-tocopherol by ascorbic acid.

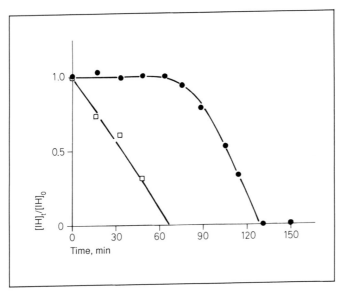

Fig. 13. Rate of consumption of antioxidant (IH) during the oxidation of 135 mM methyl linoleate at 37 °C. [AMVN] = 18.1 mM, [α-tocopherol (●)] = 0.050 mM. [5,6-C_{18}-AsA(□)] = 0.500 mM [41].

The interaction between α-tocopheroxyl radical and ascorbic acid has been established experimentally. Packer et al. [51] have shown in the pulse radiolysis study that α-tocopheroxyl radical reacts rapidly with ascorbic acid with a rate constant of k = 1.55 × 10^6 M^{-1} s^{-1}. This was also confirmed later [52] by ESR study: the ESR spectrum of α-tocopheroxyl radical disappeared quite rapidly when it was reacted with ascorbic acid. The example using fatty acid ester of ascorbic acid is shown in figure 14 [41]. Bascetta et al. [55] have also reported the results which suggest the interaction of α-tocopheroxyl radical with ascorbic acid.

The above results and discussion clearly show that ascorbic acid and α-tocopherol can act cooperatively as antioxidants in homogeneous solution. However, it has been often argued whether such an interaction can really take place in a heterogeneous system in vivo where water-soluble ascorbic acid resides in an aqueous phase while lipid-soluble α-tocopherol is incorporated into the membranes. Recent studies [32, 33] clearly show that such an interaction does take place in liposomal mem-

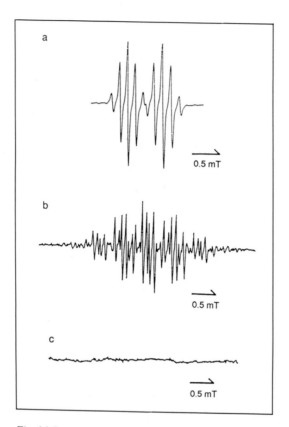

Fig. 14. Interaction of α-tocopheroxyl radical with 5,6-ascorbyl dipalmitate in benzene at 25 °C under vacuum. a ESR spectrum of galvinoxyl radical in benzene. b ESR spectrum observed when α-tocopherol was added to solution a. c ESR spectrum observed when 5,6-ascorbyl dipalmitate was added to the solution b [41].

brane systems and that ascorbic acid and α-tocopherol function synergistically.

Figure 15 shows how ascorbic acid and α-tocopherol function against the oxidations of liposomal membranes [32]. When the initial radicals are formed in the aqueous region, both ascorbic acid and α-tocopherol suppress the oxidation and produce a clear induction period. In the presence of both ascorbic acid and α-tocopherol, the length of the induction period is close to the sum of the individual induction period, that is, the effect is additive and not synergistic. Apparently, ascorbic acid traps the aqueous

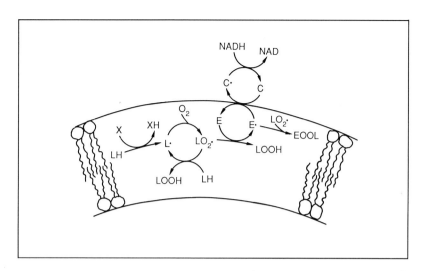

Fig. 18. Synergistic inhibition of oxidation of membranes by vitamin C and vitamin E.

It has been observed that ascorbic acid residing in the aqueous phase reduces galvinoxyl radical [59], a stable phenoxyl radical, and also nitroxide radical [60] incorporated into liposomal membranes. Ascorbic acid reduces N-oxyl-4,4′-dimethoxyloxazolidine derivatives of stearic acid (NS), well-known as spin probes, quite rapidly in the homogeneous solution. However, as shown in figure 17, when these spin probes are intercalated into the liposomal membranes, the rate of interaction between ascorbic acid and the spin probes decreases and the rate becomes slower as the nitroxide radical goes deeper into the hydrophobic phospholipid bilayer.

The efficiency or relative importance of the synergistic inhibition of oxidation by vitamin C and vitamin E is determined by several competing reactions of the vitamin E radical. Vitamin E radical may scavenge another peroxyl radical to give a stable product [61, 62], react with another vitamin E radical to give a dimer or interact with vitamin C to regenerate vitamin E (fig. 18). The regeneration of vitamin E becomes more important as the concentrations of peroxyl radical and/or vitamin E radical decrease and as the concentration of vitamin C increases. The efficiency depends also on the accessibility of vitamin C to vitamin E radical. Apparently, the efficiency is higher if the chromanoxyl radical from vitamin E is closer to the lipid-water interface, which seems to be the case indeed [63, 64].

It has not been clearly shown, however, how important this synergistic inhibition of oxidations by vitamins C and E is in biological systems. There are several reports [65–67] which suggest that such a synergism contributes even in biological systems. For example, a marginal deficiency of vitamin C was found to result in lower tocopherol levels in some tissues of guinea pigs compared to controls fed the same diet supplemented with adequate vitamin C [65]. It was also found [66] that the level of vitamin E in plasma and in lung tissue was higher in guinea pigs supplemented with ascorbic acid than in guinea pigs fed the same diet without ascorbic acid.

The vitamin C radical formed from vitamin C by a reduction of vitamin E radical may be reduced back to vitamin C by NADH-dependent system. Furthermore, dehydroascorbic acid is reduced to ascorbic acid by glutathione reductase with glutathione. Thus, a complete cycle system may be operating in biological systems to protect them against oxygen toxicity.

Tocopherols are often used as natural and safe antioxidant for foods. However, tocopherols, especially α-tocopherol, may act as prooxidants under certain conditions [50, 68–74]. It may be ascribed to the hydrogen atom abstraction from hydroperoxides contained in foods by a tocopheroxyl radical which is formed spontaneously from tocopherol. The resulting peroxyl radical can induce free radical chain oxidation of lipids in foods. When a small amount of ascorbic acid is present, however, it reduces tocopheroxyl radical and suppresses the prooxidant activity of tocopherols [73, 74].

Prooxidant Effects of Ascorbic Acid

It has been often observed that, under certain circumstances, ascorbic acid functions as a prooxidant rather than antioxidant, especially in the presence of metal ion. In fact, the combination of ascorbic acid and iron has been often used as an initiating system for free radical-mediated lipid peroxidation [13, 75]. For example, Wills [76] found that the catalytic activity of ferric ion could be strongly stimulated by the addition of ascorbic acid in the oxidation of unsaturated fatty acids. More recently, Girotti and his colleagues [77] have also observed prooxidant and antioxidant effects of ascorbate on photosensitized peroxidation of lipids in erythrocyte membranes.

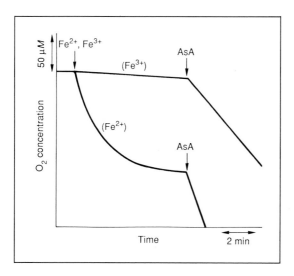

Fig. 19. Effect of addition of 100 µM ascorbic acid on the oxidation of 114 mM methyl linoleate in 10 mM Triton X-100 aqueous dispersions in the presence of 10µM iron at 37 °C [84].

The prooxidant effect of ascorbic acid does not stem from the direct attack of ascorbyl radical on the substrate to give free radicals, since the ascorbyl radical is known to be stable [19, 78, 79]. Since the first observation by Yamazaki et al. [80] of the ascorbyl radical by ESR, a number of ESR investigations have been carried out [46, 78, 79, 81] and it is now accepted that the unpaired electron of ascorbyl radical is spread over a highly conjugated tricarbonyl system [78]. The delocalized nature of the unpaired electron in the ascorbyl radical makes it relatively unreactive.

Ascorbic acid is a strong reducing agent and it reduces ferric ion (Fe^{3+}) to ferrous ion (Fe^{2+}). Ferrous ion decomposes hydrogen peroxide and hydroperoxide much faster than ferric ion [82–84]. Thus ascorbic acid produces more active ferrous ion from less active ferric ion.

The spontaneous oxidation of methyl linoleate micelles in aqueous dispersions was quite small, but the addition of ferrous salt into the aqueous phase induced the oxidation without any appreciable induction period (fig. 19) [83, 85]. The rate of oxidation increased with increasing concentration of ferrous salt, the kinetic order of the initial rate of oxidation on ferrous ion concentration being about 0.5. When the surface charge of the micelles was either neutral or negative, ferrous ion induced the oxidation, but it could not induce the oxidation when the micelles had a positive charge in their surface. Even in this case, however, the addition of tert-butyl hydroperoxide to the aqueous phase induced a fast oxidation without induction period. Apparently, ferrous ion decomposes lipid hydroperoxide present concomitantly in the micelles to generate lipid alkoxyl radical, which induced the chain oxidation. When the surface charge of the micelles is positive, the ferrous ion may not be able to access to the hydroperoxide, although neutral tert-butoxyl radical formed in the aqueous phase by the interaction of ferrous ion and tert-butyl hydroperoxide can enter the micelles and induce chain oxidation independent of the surface charge of the micelles.

Ferric ion also induced the oxidation of methyl linoleate micelles, but the rate of oxidation was much smaller than that induced by ferrous ion [83, 85]. In the presence of both ferrous and ferric ions, the rate of the oxidation increased monotonously with increasing ratio of ferrous ion [85]. The addition of ascorbic acid enhanced the oxidation markedly in the presence of ferric ion (fig. 19) [83], apparently because ascorbic acid reduced ferric ion to more active ferrous ion. As shown in figure 19, the rate of oxidation of methyl linoleate micelles induced by ferrous ion decreased with time, but the addition of ascorbic acid resumed a fast oxidation. This is because ferrous ion oxidized to ferric ion as the oxidation proceeds is reduced to ferrous ion again by ascorbic acid.

Thus, ascorbic acid may function as a prooxidant in the presence of ferric ion by reducing ferric ion to ferrous ion which decomposes peroxides much faster than ferric ion to generate oxygen radicals. Ascorbic acid is consumed by the reaction with ferric ion, but if excess amount of ascorbic acid is present, it acts as an antioxidant. Figure 20 shows that the excess ascorbic acid suppresses the oxidation of methyl linoleate micelles in aqueous dispersions induced by ferrous sulfate.

Similar effects of ascorbic acid have been also reported: for example, Rees and Slater [86] observed a prooxidant effect of ascorbic acid at low concentrations and an antioxidant effect at high concentration in the NADPH-dependent, NADPH-CCl_4-dependent, and cumene hydroperoxide-dependent lipid peroxidations in rat liver microsomal suspensions.

As mentioned above, ascorbic acid is autoxidized to give superoxide and hydrogen peroxide. This is not a fast reaction, but under certain conditions, they may contribute for the prooxidant effect of ascorbic acid.

It is clear from the above discussion that ascorbic acid can function as both antioxidant and prooxidant. Ascorbic acid, like oxygen and free rad-

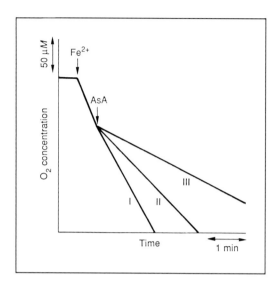

Fig. 20. Effect of addition of ascorbic acid (I: 10 μ*M*, II: 100 μ*M*, III: 1 m*M*) on the oxidation of 144 m*M* methyl linoleate in 10 m*M* Triton X-100 aqueous dispersions in the presence of 10 μ*M* ferrous sulfate at 37 °C [84].

icals, is a double edged sword and its overall effect is determined by a delicate balance of many factors. Although ascorbic acid acts as a potent radical scavenger and suppresses the in vitro oxidations of methyl linoleate micelles and soybean PC liposomes in aqueous dispersions and also the oxidative hemolysis of erythrocytes induced by AAPH, its role in the AAPH-induced intoxication for mouse is complicated [87]. However, there is little evidence that the prooxidant activity is of any significance in vivo [8]. On the other hand, when the radicals are generated in the plasma vitamin C is consumed faster than uric acid, vitamin E, bilirubin and thiols, suggesting that vitamin C acts as a primary defense in the plasma.

Controversy about biological functions such as toxic effect by megadose, protection against cancer, and effects on aging persists. Some publications report the beneficial effect in a number of clinical complaints including the common cold, asthma, atherosclerosis and cancer, but in many instances scientific data supporting these claims are lacking. Apparently, more work is necessary to fully understand and elucidate the role of vitamin C in biological systems.

References

1 Lewin, S.: Vitamin C: its molecular biology and medical potential (Academic Press, London 1976).

2 Counsel, J.N.; Hornig, D.H.: Vitamin C: ascorbic acid. (Applied Sci. Pub., London 1981).

3 Tolbert, B.M.: Metabolism and function of ascorbic acid and its metabolites. Int. J. Vitam. Nutr. Res. *27:* 122–138 (1985).

4 Levine, M.: New concepts in the biology and biochemistry of ascorbic acid. New Engl. J. Med. *314:* 892–901 (1986).

5 Seib, P.A.; Tolbert, B.M. (eds.): Ascorbic acid: chemistry, metabolism, and uses. Adv. Chem. Ser. *200:* 1–604 (1982).

6 Burns, J.J.; Rivers, J.M.; Machlin, L.J. (eds.): Third Conference on Vitamin C. Ann. N.Y. Acad. Sci. *498:* 1–538 (1987).

7 Cort, W.M.: Antioxidant properties of ascorbic acid in foods. Adv. Chem. Ser. *200:* 533–550 (1982).

8 Bendich, A.; Machlin, L.J.; Scandurra, O.; Burton, G.W.; Wayner, D.D.M.: The antioxidant role of vitamin C. Adv. Free Radical Biol. Med. *2:* 419–444 (1986).

9 Matusch, R.Z.: Reductones and reductonates. I. The redox pair dehydroascorbic acid-ascorbic acid and anions. Z. Naturforsch., B, Anorg. Chem. *32B:* 562–568 (1977).

10 Porter, N.A.: Mechanisms for the autoxidation of polyunsaturated lipids. Acc. Chem. Res. *19:* 262–268 (1986).

11 Pryor, W.A.: Free radical biology: xenobiotics, cancer, and aging. Ann. N.Y. Acad. Sci. *393:* 1–22 (1982).

12 Hayaishi, O.; Niki, E.; Kondo, M.; Yoshikawa, T. (eds.): Medical, biochemical, and chemical aspects of free radicals (Elsevier, Amsterdam 1988).

13 Halliwell, B.; Gutteridge, J.M.C.: Free radicals in biology and medicine (Clarendon Press, Oxford 1982).

14 Sies, H. (ed.): Oxidative stress (Academic Press, London 1985).

15 Niki, E.: Antioxidants in relation to lipid peroxidation. Chem. Phys. Lipids *44:* 227–253 (1987).

16 Burton, G.W.; Foster, D.O.; Perly, B.; Slater, T.F.; Smith, I.C.P.; Ingold, K.U.: Biological antioxidant. Phil. Trans. R. Soc. Lond. *B311:* 565–578 (1985).

17 Nishikimi, M.: Oxidation of ascorbic acid with superoxide anion generated by the xanthine-xanthine oxidase system. Biochem. biophys. Res. Commun. *63:* 463–468 (1975).

18 Cabelli, D.E.; Bielski, B.H.J.: Kinetics and mechanism for the oxidation of ascorbic acid/ascorbate by HO_2/O_2^- radicals. A pulse radiolysis and stopped-flow photolysis study. J. phys. Chem. *87:* 1809–1812 (1983).

19 Bielski, B.H.J.: Chemistry of ascorbic acid radicals. Adv. Chem. Ser. *200:* 81–100 (1982).

20 Packer, J.E.; Willson, R.L.; Bahnemann, D.; Asmus, K.-D.: Electron transfer reactions of halogenated aliphatic peroxyl radicals: measurement of absolute rate constants by pulse radiolysis. J.C.S. Perkin II 296–299 (1980).

21 Bielski, B.H.J.; Cabelli, D.E.; Arudi, R.L.; Ross, A.B.: Reactivity of HO_2/O_2^- radicals in aqueous solution. J. phys. Chem. Ref. Data. *14:* 1041–1100 (1985).

22   Sawyer, D.T.; Chiericato, G. Jr.; Tsuchiya, T.: Oxidation of ascorbic acid and dehy-droascorbic acid by superoxide ion in aprotic media. J. Am. chem. Soc. *104:* 6273–6278 (1982).

23   Chon, P-T.; Khan, A.U.: *L*-ascorbic acid quenching of singlet delta molecular oxy-gen in aqueous media: Generalized antioxidant property of vitamin C. Biochem. Biophys. Res. Commun. *115:* 932–937 (1983).

24   Yamamoto, Y.; Haga, S.; Niki, E.; Kamiya, Y.: Oxidation of lipids. V. Oxidation of methyl linoleate in aqueous dispersion. Bull. Chem. Soc. Japan *57:* 1260–1264 (1984).

25   Barclay, L.R.C.; Locke, S.J.; MacNeil, J.M.; Van Kessel, J.; Burton, G.W.; Ingold, K.U.: Autoxidation of micelles and model membranes. Quantitative kinetic mea-surements can be made by using either water-soluble or lipid-soluble initiators with water-soluble or lipid-soluble chain-breaking antioxidants. J. Am. chem. Soc. *106:* 2479–2481 (1984).

26   Howard, J.A.; Ingold, K.U.: Absolute rate constants for hydrocarbon autoxidation. 6. Aromatics and olefins. Can. J. Chem. *45:* 793–802 (1967).

27   Yamamoto, Y.; Niki, E.; Kamiya, Y.: Oxidation of lipids. III. Oxidation of methyl linoleate in solution. Lipids *17:* 870–877 (1982).

28   Niki, E.; Saito, T.; Kawakami, A.; Kamiya, Y.: Inhibition of oxidation of methyl linoleate in solution by vitamin E and vitamin C. J. biol. Chem. *259:* 4177–4182 (1984).

29   Niki, E.; Yamamoto, Y.; Kamiya, Y.: Oxidation of phosphatidylcholine and its inhibition by vitamin E and vitamin C; in Bors, Saran, Tait, Oxygen radicals in chemistry and biology, pp. 273–280 (de Gruyter, Berlin 1984).

30   Wayner, D.D.M.; Burton, G.W.; Ingold, K.U.: The antioxidant efficiency of vitamin C is concentration-dependent. Biochim. biophys. Acta *884:* 119–123 (1986).

31   Scarpa, M.; Stevanato, R.; Viglino, P.; Rigo, A.: Superoxide ion as active interme-diate in the autoxidation of ascorbate by molecular oxygen. J. biol. Chem. *258:* 6695–6697 (1983).

32   Niki, E.; Kawakami, A.; Yamamoto, Y.; Kamiya, Y.: Oxidation of lipids. VIII. Synergistic inhibition of oxidation of vitamin E and vitamin C. Bull. chem. Soc. Japan *58:* 1971–1975 (1985).

33   Doba, T.; Burton, G.W.; Ingold, K.U.: Antioxidant and co-antioxidant activity of vitamin C. The effect of vitamin C, either alone or in the presence of vitamin E or a water-soluble vitamin E analogue, upon the peroxidation of aqueous multilamellar phospholipid liposomes. Biochim. biophys. Acta *835:* 298–303 (1985).

34   Yamamoto, Y.; Niki, E.; Kamiya, Y.; Miki, M.; Tamai, H.; Mino, M.: Free radical chain oxidation and hemolysis of erythrocytes by molecular oxygen and their inhi-bition by vitamin E. J. Nutr. Sci. Vitaminol. *32:* 475–479 (1986).

35   Miki, M.; Tamai, H.; Mino, M.; Yamamoto, Y.; Niki, E.: Free-radical chain oxida-tion of rat red blood cells by molecular oxygen and its inhibition by α-tocopherol. Archs Biochem. Biophys. *258:* 373–380 (1987).

36   Wayner, D.D.M.; Burton, G.W.; Ingold, K.U.; Locke, S.: Quantitative measurement of the total, peroxyl radical-trapping antioxidant capability of human blood plasma by controlled peroxidation. The important contribution made by plasma proteins. FEBS Lett. *187:* 33–37 (1985).

37   Wayner, D.D.M.; Burton, G.W.; Ingold, K.U.; Barkley, L.R.C.; Locke, S.J.: Bio-chim. biophys. Acta *924:* 408–419 (1987).

38 Niki, E.; Yamamoto, Y.; Takahashi, M.; Yamamoto, K.; Yamamoto, Y.; Komuro, E.; Miki, M.; Yasuda, H.; Mino, M.: Free radical-mediated damage of blood and its inhibition by antioxidants. Vitamins, Kyoto *62:* 200 (1988).

39 Kunert, K.-J.; Tappel, A.L.: The effect of vitamin C on in vivo lipid peroxidation in guinea pigs as measured by pentane and ethane production. Lipids *18:* 271–294 (1983).

40 Kato, N.; Kawai, K.; Yoshida, A.: Effect of dietary level of ascorbic acid on the growth, hepatic lipid peroxidation, and serum lipids in guinea pigs fed polychlorinated biphenyls. J. Nutr. *111:* 1727–1733 (1981).

41 Takahashi, M.; Niki, E.; Kawakami, A.; Kumasaka, A.; Yamamoto, Y.; Kamiya, Y.; Tanaka, K.: Oxidation of lipids. XIV. Inhibition of oxidation of methyl linoleate by fatty acid esters of L-ascorbic acid. Bull. chem. Soc. Japan *59:* 3179–3183 (1986).

42 Gwo, Y.-Y.; Flick, Jr., G.J.; Dupuy, H.P.: Effect of ascorbyl palmitate on the quality of frying fats for deep frying operations. J. Am. Oil Chem. Soc. *62:* 1666–1671 (1985).

43 Niki, E.: Interaction of ascorbate and α-tocopherol. Ann. N. Y. Acad. Sci. *498:* 186–199 (1987).

44 Burton, G.W.; Ingold, K.U.: Vitamin E: Application of the principles of physical organic chemistry to the exploration of its structure and function. Acc. Chem. Res. *19:* 194–201 (1986).

45 McCay, P.B.: Vitamin E: Interactions with free radicals and ascorbate. Ann. Rev. Nutr. *5:* 323–340 (1985).

46 Craw, M.T.; Depew, M.C.: Contributions of electron spin responance spectroscopy to the study of vitamins C, E and K. Rev. Chem. Intermed. *6:* 1–31 (1985).

47 Golumbic, C.; Mattill, H.A.: Antioxidants and the autoxidation of fats. XIII. The antioxygenic action of ascorbic acid in association with tocopherols, hydroquinones and related compounds. J. Am. chem. Soc. *63:* 1279–1280 (1941).

48 Tappel, A.L.: Will antioxidant nutrients slow aging processes? Geriatrics *23:* 97–105 (1968).

49 Cort, W.M.: Antioxidant activity of tocopherols, ascorbyl palmitate, and ascorbic acid and their mode of action. J. Am. Oil Chem. Soc. *51:* 321–325 (1974).

50 Chen, L.H.; Chang, M.L.: Effect of dietary vitamin E and Vitamin C on respiration and swelling of guinea pig liver mitochondria. J. Nutr. *108:* 1616–1620 (1978).

51 Packer, J.E.; Slater, T.F.; Willson, R.L.: Direct observation of a free radical interaction between vitamin E and vitamin C. Nature, Lond. *278:* 737–738 (1979).

52 Niki, E.; Tsuchiya, J.; Tanimura, R.; Kamiya, Y.: Regeneration of vitamin E from α-chromanoxy radical by glutathione and vitamin C. Chem. Let. 789–792 (1982).

53 Barclay, L.R.C.; Locke, S.L.; MacNeil, J.M.: The autoxidation of unsaturated lipids in micelles. Synergism of inhibitors vitamins C and E. Can. J. Chem. *61:* 1288–1290 (1983).

54 Barclay, L.R.C.: Autoxidation in micelles. Synergism of vitamin C with lipid-soluble vitamin E and water-soluble Trolox. Can. J. Chem. *63:* 366–374 (1985).

55 Bascetta, E.; Gunstone, F.E.; Walton, J.C.: Electron spin resonance study of the role of vitamin E and vitamin C in the inhibition of fatty acid oxidation in a model membrane. Chem. Phys. Lipids *33:* 207–210 (1983).

56   Scarpa, M.; Rigo, A.; Maiorino, M.; Ursini, F.; Gregolin, C.: Formation of α-toco-
      pherol radical and recycling of α-tocopherol by ascorbate during peroxidation of
      phosphatidylcholine liposomes. Biochim. biophys. Acta 801: 215–219 (1984).

57   Liebler, D.C.; Kling, D.S.; Reed, D.J.: Antioxidant protection of phospholipid
      bilayers by α-tocopherol. J. biol. Chem. 261: 12114–12119 (1986).

58   Leung, H.-W.; Vang, M.J.; Mavis, R.D.: The cooperative interaction between vita-
      min E and vitamin C in suppression of peroxidation of membrane phospholipids.
      Biochim. biophys. Acta 664: 266–272 (1981).

59   Tsuchiya, J.; Yamada, T.; Niki, E.; Kamiya, Y.: Interaction of galvinoxyl radical
      with ascorbic acid, cysteine, and glutathione in homogeneous solution and in
      aqueous dispersions. Bull. chem. Soc. Japan 58: 326–330 (1985).

60   Takahashi, M.; Tsuchiya, J.; Niki, E.; Urano, S.: Action of vitamin E as antioxidant
      in phospholipid liposomal membranes as studied by spin label technique. J. Nutr.
      Sci. Vitaminol. 34: 25–34 (1988).

61   Winterle, J.; Dulin, D.; Mill, T.: Products and stoichiometry of reaction of vitamin E
      with alkylperoxy radicals. J. org. Chem. 49: 491–495 (1984).

62   Matsumoto, S.; Matsuo, M.; Iitaka, Y.; Niki, E.: Oxidation of a vitamin E model
      compound, 2,2,5,7,8-pentamethylchroman-6-ol, with the α-tocopheroxyl radical. J.
      chem. Soc., chem. Commun. 1076–1077 (1986).

63   Perly, B.; Smith, J.C.P.; Hughes, L.; Burton, G.W.; Ingold, K.U.: Estimation of the
      location of natural alpha-tocopherol in lipid bilayers by C-NMR spectroscopy. Bio-
      chim. biophys. Acta 819: 131–135 (1985).

64   Kagan, V.E.; Serbinova, E.A.; Bakalova, R.A.; Novikov, K.N.; Skrypin, V.I.; Evstig-
      neeva, R.P.; Stoytchev, Ts.S.: Effects of alpha-tocopherol derivatives with different
      chain length on in vitro and in vivo lipid peroxidation in liver microsomes; in
      Rice-Evans, Free radicals, oxidant stress and drug action, pp. 425–442 (1987).

65   Hruba, F.; Novakova, V.; Ginter, E.: The effect of chronic marginal vitamin C
      deficiency on the alpha-tocopherol content of the organs and plasma of guinea-pigs.
      Experientia 38: 1454–1455 (1982).

66   Bendich, A.; D'Apolito, P.; Gabriel, E.; Machlin, L.J.: Interaction of dietary vitamin
      C and vitamin E on guinea pig immune responses to mitogens. J. Nutr. 114: 1588–
      1593 (1984).

67   Miyazawa, T.; Ando, T.; Kaneda, T.: Effect of dietary vitamin C and vitamin E on
      tissue lipid peroxidation of guinea pigs fed with oxidized oil. Agric. Biol. Chem. 50:
      71–78 (1986).

68   Kanno, C.; Yamauchi, K.; Tsugo, T.: Antioxidant effect of tocopherols on autoxida-
      tion of milk fat. Agric. Biol. Chem. 34: 886–890 (1970).

69   Cillard, J.; Cillard, P.: Behavior of alpha, gamma, and delta tocopherols with linoleic
      acid in aqueous media. J. Am. Oil Chem. Soc. 57: 39–42 (1980).

70   Cillard, J.; Cillard, P.; Cormier, M.: α-Tocopherol prooxidant effect in aqueous
      media: Increased autoxidation rate of linoleic acid. J. Am. Oil Chem. Soc. 57: 252–
      255 (1980).

71   Cillard, J.; Cillard, P.; Cormier, M.: Effect of experimental factors on the prooxidant
      behaviour of α-tocopherol. J. Am. Oil Chem. Soc. 57: 255–261 (1980).

72   Bazin, B.; Cillard, J.; Koskas, J.-P.; Cillard, P.: Arachidonic acid autoxidation in an
      aqueous media. Effect of α-tocopherol, cysteine, and nucleic acids. J. Am. Oil Chem.
      Soc. 61: 1212–1215 (1984).

73 Terao, J.; Matsushita, S.: The peroxidizing effect of α-tocopherol on autoxidation of methyl linoleate in bulk phase. Lipids *21:* 255–260 (1986).

74 Takahashi, M.; Niki, E.; Yoshikawa, Y.; Kamiya, Y.: Prooxidant effect of α-tocopherol. 24th General Meeting of Japanese Oil Chemists' Society, October, 1985.

75 Cheeseman, K.H.; Burton, G.W.; Ingold, K.U.; Slater, T.F.: Lipid peroxidation and lipid antioxidants in normal and tumor cells. Toxicol. Pathol. *12:* 235–239 (1984).

76 Wills, E.D.: Mechanisms of lipid peroxide formation in tissues. Role of metals and haematin proteins in the catalysis of the oxidation of unsaturated fatty acids. Biochim. biophys. Acta *98:* 238–251 (1965).

77 Girotti, A.W.; Thomas, J.P.; Jordan, J.E.: Prooxidant and antioxidant effects of ascorbate on photosensitized peroxidation of lipids in erythrocyte membranes. Photochem. Photobiol. *41:* 267–276 (1985).

78 Laroff, G.P.; Fessenden, R.W.; Schuler, R.H.: The electron spin resonance spectra of radical intermediates in the oxidation of ascorbic acid and related substances. J. Am. chem. Soc. *94:* 9062–9073 (1972).

79 Swartz, H.M.; Dodd, N.J.F.: The role of ascorbic acid on radical reaction in vivo; in Rodgers, Oxygen and oxygen radicals in chemistry and biology, pp. 161–168 (1981).

80 Yamazaki, I.; Mason, H.S.; Piette, L.: Identification by electron paramagnetic resonance spectroscopy of free radicals generated from substrates by peroxides. J. biol. Chem. *235:* 2444–2449 (1960).

81 Kirino, Y.; Kwan, T.: Free radicals formed during the oxidation of *L*-ascorbic acid or hydroxytetronic acid with hydrogen peroxide and titanium(III) ions. Chem. pharm. Bull., Tokyo *19:* 718–721 (1971).

82 Walling, C.: Fenton's reagent revisited. Acc. Chem. Res. *8:* 125–131 (1975).

83 Yamamoto, K.; Takahashi, M.; Niki, E.: Role of iron and ascorbic acid in the oxidation of methyl linoleate micelles. Chem. Lett. 1149–1152 (1987).

84 Yamamoto, K.; Niki, E.: Interaction of α-tocopherol with iron: antioxidant and prooxidant effects of α-tocopherol in the oxidation of lipids in aqueous dispersions in the presence of iron. Biochim. biophys. Acta *958:* 19–23 (1988).

85 Niki, E.; Yamamoto, K.; Takahashi, M.: Role of iron, ascorbic acid and tocopherol in the oxidation of lipids; in Ando, Morooka, The role of oxygen in chemistry and biochemistry, pp. 509–514 (Elsevier, Amsterdam 1988).

86 Rees, S.; Slater, T.F.: Ascorbic acid and lipid peroxidation: The cross-over effect. Acta biochim. biophys. Hung. *22:* 241–249 (1987).

87 Terao, K.; Niki, E.: Damage to biological tissues induced by radical initiator 2,2'-azobis(2-amidinopropane) dihydrochloride and its inhibition by chain-breaking antioxidants. J. Free Rad. Biol. Med. *2:* 193–201 (1986).

Etsuo Niki, PhD, Department of Reaction Chemistry, Faculty of Engineering, The University of Tokyo, Hongo, Tokyo 113 (Japan)

Simopoulos AP (ed): Selected Vitamins, Minerals, and Functional Consequences of
Maternal Malnutrition. World Rev Nutr Diet. Basel, Karger, 1991, vol 64, pp 31–57

Smoking and Vitamin C

Akira Murata

Department of Applied Biological Sciences, Saga University, Saga, Japan

Contents

Cigarette smoking is currently a serious public health problem of
increasing concern worldwide. Cigarette smoke contains a large number of
toxic agents and its chronic inhalation gives rise to adverse effects on
numerous physiological and biochemical functions. Based on vast
amounts of epidemiological, clinical and experimental evidence, it is generally accepted that long-term cigarette smoking is associated with an

increased risk of lung and larynx cancer, several other cancers, cardiovascular disease, cerebrovascular disease, chronic obstructive pulmonary disease and several other chronic diseases [90, 91].

Vitamin C, in addition to its fundamental action as the anti-scorbutic vitamin, has been found to be an essential factor in numerous physiological, biological and biochemical functions of metabolic and clinical importance. As a consequence, vitamin C supplementation has been used by millions of people throughout the world, although beneficial roles to health by vitamin C supplementation are still controversial.

In numerous studies, lower vitamin C levels in whole blood, serum, plasma, leukocytes, urine etc. have been found in cigarette smokers when compared with nonsmokers. It has been pointed out that several smoking-related diseases are associated with lowered vitamin C levels [13, 42]. On the other hand, the vitamin C levels of cigar or pipe smokers appear to be comparable to those of nonsmokers. Cigar and pipe smoking are associated with a lower risk of most smoking-related diseases.

It has been pointed out that cigarette smokers have an increased risk of marginal vitamin C deficiency and its health consequences. Cigarette smokers have a special requirement of vitamin C to compensate for their lowered levels. Daily vitamin C supplement, in addition to proper diet, would assure satisfactory levels of vitamin C, thus lessening the risk of marginal vitamin C deficiency and its health consequences. Hornig and Glatthaar [42] have reviewed the literature concerned with vitamin C and smoking up to 1982 and discussed mainly the requirement of vitamin C in smokers.

In this article I present conclusive evidence of lowered vitamin C levels in smokers, summarize the available literature on smoking and vitamin C and discuss the effects of supplementary vitamin C. Several more general books and reviews on vitamin C are included in the bibliography [6, 9, 13, 15, 18, 23, 24, 26, 34, 43, 48, 60, 62, 63, 79]. Smoking and smokers hereinafter mean cigarette smoking and cigarette smokers, respectively, unless otherwise stated.

Lowered Vitamin C Levels of Smokers

Already in 1941, low vitamin C levels in the blood of heavy smokers were reported. There were several reports suggesting lowered vitamin C levels in smokers during the period from the 1940s to the 1960s, although

Table 1. Whole blood vitamin C levels of smokers and nonsmokers

Sex	Vitamin C, mg/dl				p	Reference
	smokers	n	nonsmokers	n		
M	0.30 ± 0.09	5	0.66 ± 0.21	5	<0.01	71
M + F	0.42 ± 0.06	14	0.60 ± 0.05	14	<0.05	72

there were also contradictory reports (see Hornig and Glatthaar [42] for details and references). In this article the studies after 1970 will be reviewed, including several early important reports. The data from those taking vitamin C supplement will be excluded, unless otherwise stated.

Whole Blood

Pelletier [71, 72] demonstrated that whole blood vitamin C levels of smokers were significantly lower than those of nonsmokers (table 1) although the smokers and nonsmokers had similar characteristics and dietary intakes of vitamin C.

Serum

The nutritional status of vitamin C is usually based upon the determination of serum or plasma levels of the vitamin. Even though the influence of recent vitamin C intake on blood levels is widely recognized, serum and plasma are used because of the ready availability of the specimen.

Pelletier [73, 74] examined the serum vitamin C levels of smokers and nonsmokers from various regions of Canada, using data from the Nutrition Canada National Survey conducted from 1970 to 1972. The median serum vitamin C levels for both males and females aged 20–64 years were reduced by about 25% in smokers of less than 20 cigarettes a day and by about 40% in smokers of 20 cigarettes or more a day in comparison to nonsmokers. In Quebec region, more than 50% of the male smokers were classified as being at risk according to the Nutrition Canada interpretive guidelines for high (<0.2 mg/dl) and moderate (0.2–0.4 mg/dl) risk [45], as compared to less than 15% of the nonsmokers.

Smith and Hodges [81] analyzed data from the second National Health and Nutrition Examination Survey of the United States conducted from 1976 to 1980, and found that the mean serum vitamin C levels of

Table 2. Serum vitamin C levels of smokers and nonsmokers (adapted from Schectman et al. [77])

Group	n	Vitamin C, mg/dl	
		mean	95% CI
Smokers			
> 20 cigarettes/day	1,288	0.83	(0.79, 0.88)
20 cigarettes/day	1,454	0.82	(0.77, 0.86)
< 20 cigarettes/day	1,553	0.97	(0.92, 1.03)
Nonsmokers			
Stopped smoking within the past year	346	0.99	(0.92, 1.06)
Stopped for the previous year or longer	1,942	1.10	(1.06, 1.14)
Never smoked	5,009	1.15	(1.15, 1.18)

smokers were approximately 0.2 mg/dl lower than those of nonsmokers having similar dietary intakes of vitamin C. The percentage of smokers with serum vitamin C levels of 0.3 mg/dl or less were 2–4 times greater than in nonsmokers.

More recently, analyzing data from the second National Health and Nutrition Examination Survey of the United States, Schectman et al. [77] confirmed that smoking was associated with lowered serum vitamin C levels (table 2). They also found that the inverse association between serum vitamin C levels and smoking was independent of age, sex, body weight, race, and alcoholic beverage consumption.

Plasma

There have been numerous reports on plasma vitamin C levels of smokers (table 3).

In the 1960s, Calder et al. [17], Pelletier [71], and Brook and Grimshaw [14] found lower levels of vitamin C in the plasma of smokers than of nonsmokers. Calder et al. [17] showed that the plasma vitamin C levels of smokers consuming 15 cigarettes or more daily were lower than of smokers consuming less than 15 cigarettes daily. Brook and Grimshaw [14], in a study of males and females aged 17 to 63 years, reported that the plasma vitamin C levels of smokers were lower than those of nonsmokers, and also observed that the plasma vitamin C levels of smokers were lower in males than in females.

Table 3. Plasma vitamin C levels of smokers and nonsmokers

Sex	Vitamin C, mg/dl				p	Reference
	smokers	n	nonsmokers	n		
–	0.73 ± 0.04	83	0.91 ± 0.05	91	< 0.005	17
–	0.52 ± 0.07	31	0.91 ± 0.05	91	< 0.001	17
M	0.29 ± 0.08	5	0.72 ± 0.29	5	< 0.05	71
M	0.44 ± 0.06	22	0.62 ± 0.07	32	< 0.001	14
F	0.74 ± 0.07	34	0.97 ± 0.05	50	< 0.001	14
F	0.54 ± 0.07	61	0.63 ± 0.03	154	< 0.001	25
F	0.45 ± 0.08	39	0.63 ± 0.03	154	< 0.001	25
M	0.61	20	0.66	20	–	4
M	0.16	40	0.20	39	–	16
M	0.27	9	0.37	21	–	16
M + F	0.13 ± 0.02	18	0.18 ± 0.01	34	< 0.001	1
M	0.51 ± 0.37	96	0.78 ± 0.38	80	< 0.01	64
M	0.93	–	0.97	–	–	38
F	0.74	–	0.95	–	–	38
M	0.50	–	0.58	–	–	38
F	0.56	–	0.63	–	–	38
M	0.69 ± 0.26	12	0.88 ± 0.27	10	< 0.05	53
M	0.73 ± 0.42	8	1.44 ± 0.45	20	< 0.01	10
M	0.53 ± 0.41	8	1.12 ± 0.61	20	< 0.01	10
M	0.39 ± 0.13	96	0.66 ± 0.21	100	< 0.001	67
M	0.56 ± 0.04	52	0.87 ± 0.03	66	< 0.001	22
M	0.54 ± 0.26	24	0.73 ± 0.30	59	< 0.01	8
F	0.32 ± 0.19	11	1.46 ± 0.69	26	< 0.001	56
F	0.40 ± 0.32	14	0.89 ± 0.27	12	< 0.01	69
F	0.51 ± 0.24	46	0.60 ± 0.26	103	< 0.05	7
M	0.49 ± 0.15	51	0.72 ± 0.23	73	< 0.001	68
M	0.56 ± 0.13	48	0.84 ± 0.21	88	< 0.001	68
M	0.52 ± 0.15	56	0.86 ± 0.24	93	< 0.001	68

Elwood et al. [25] showed that among female smokers plasma vitamin C levels of those smoking 15 cigarettes or more daily were lower than of those smoking less than 15 cigarettes daily. Burr et al. [16] examined the plasma vitamin C levels in the elderly aged 75 years and over, and found lower levels in smokers than in nonsmokers even though the mean values were in the subnormal range. Albanese et al. [1] reported lower levels of vitamin C in smokers than in nonsmokers, while the mean values were also in the subnormal range. McClean et al. [64] studied the plasma vitamin C levels of smokers in relation to age. They confirmed significantly lower levels in smokers aged 17 to 39 years, but found that as age increased the difference between smokers' and nonsmokers' plasma vitamin C levels became less, in accord with facts that vitamin C levels fall with increasing age.

Hoefel [38] analyzed 712 subjects of both sexes for vitamin C level, dividing the subjects into groups in accordance with their social and economic conditions. He observed that the plasma vitamin C levels were lower in smokers, and the difference was more marked in female subjects. In the lower socio-economic classes, average plasma vitamin C levels were within the range known as high risk, i.e. under 0.2 mg/dl and subclinical signs of vitamin C deficiency were noted. Ritzel and Bruppacher [76] measured the plasma vitamin C levels in 816 smoking and 1058 nonsmoking males aged 20 to 75 of the Basel metropolitan area. The plasma vitamin C levels were lower in smokers than in nonsmokers. About 10% of the nonsmokers were at risk according to the Canadian criteria [45]. This percentage increased gradually with increasing number of cigarettes to about 40% in smokers consuming more than 30 cigarettes daily. According to the US standards, 9% of the heavy smokers had deficient vitamin C levels, while the proportion of nonsmokers was less than 2%. Similar findings were obtained from 218 smoking and 432 nonsmoking females.

Keith and Driskel [53] showed that plasma vitamin C levels were significantly lower in smokers, although dietary vitamin C intakes for smokers and nonsmokers were not significantly different. Biersner et al. [10] determined the plasma vitamin C levels of crew members of a ballistic missile nuclear submarine prior to, during and after a long patrol. They found that during the late-patrol period plasma levels were lower than pre- and post-patrol, and in all periods smokers had significantly lower levels of plasma vitamin C than nonsmokers.

We [67, 68] examined the vitamin C status of smokers and nonsmokers of Japanese over a 4-year period, and confirmed the findings of earlier reports based on Caucasians. The data from 1985 are shown in figure 1. The smokers had lower plasma vitamin C levels than the nonsmokers with similar vegetable and fruit consumption. The inverse association between smoking and vitamin C levels was independent of alcoholic beverage consumption, although alcohol consumption was greater in smokers than in nonsmokers. The percentages of smokers classified as being at risk according to Canadian criteria [45] were more than 10 times greater than those of nonsmokers. Chow et al. [22] determined the plasma levels of micronu-

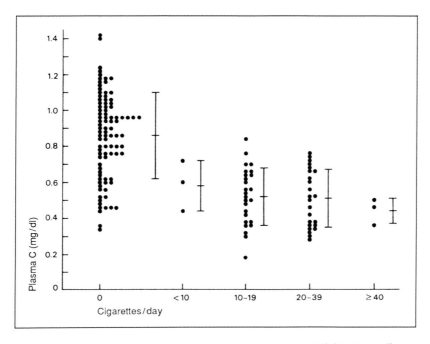

Fig. 1. Plasma vitamin C levels of smokers and nonsmokers subdivided according to daily cigarette consumption. Murata et al. [68]. Bars represent mean ± SD.

trients, and found that plasma levels of vitamin C and total carotenes were significantly lower in smokers than in nonsmokers, while plasma levels of vitamin A, selenium and vitamin E were not significantly different between the two groups.

The nutritional status of adolescent females aged 14 to 17 years was evaluated with respect to vitamin C [56]. The smokers had lower dietary vitamin C intakes and plasma vitamin C levels than nonsmokers. When plasma vitamin C values were adjusted for vitamin C intakes, smokers still exhibited significantly lower plasma vitamin C levels. Norkus et al. [69] examined the vitamin C levels of pregnant women at delivery, and found that the smokers had significantly lower vitamin C levels in maternal and umbilical cord blood plasma than nonsmokers. More recently, Basu et al. [7] examined plasma vitamin C levels in women who smoked and were on oral contraception. Significant decreases in plasma vitamin C levels were observed in smokers. No association between smoking and oral contracep-

Table 4. Leukocyte vitamin C concentrations of smokers and nonsmokers

Sex	Vitamin C, $\mu g/10^8$ cells				p	Reference
	smokers	n	nonsmokers	n		
–	24.3 ± 0.08	83	29.1 ± 0.9	91	< 0.005	17
–	21.7 ± 1.6	31	29.1 ± 0.9	91	< 0.001	17
M	18.8 ± 1.6	22	24.6 ± 1.3	32	< 0.05	14
F	26.0 ± 1.3	34	30.7 ± 1.4	50	< 0.05	14
M	15.7	29	17.5	60	–	16
M	13.8	20	17.5	60	–	16
M + F	18.9	18	29.1	34	–	1
M	38.8 ± 12.1	23	50.0 ± 20.5	18	< 0.05	57
M	13.5 ± 6.6	98	19.9 ± 7.4	80	< 0.001	64
F	30.2 ± 11.4	243	36.1 ± 12.8	760	< 0.001	78
F	28.3 ± 11.8	15	36.1 ± 12.8	760	< 0.05	78

tive use was seen for vitamin C levels among women less than 26 years old, but decreases in vitamin C levels were evident among smokers aged 26 years or over. In general, the blood vitamin C levels are lower in males than in females. This sex difference still persists in smoking males and females.

Effects of smoking on plasma vitamin C levels were studied in rats [19]. Three- or 7-day exposure to cigarette smoke resulted in significant decreases in vitamin C levels in the plasma. Unlike humans, rats are capable of synthesizing vitamin C for their needs. The lowering effect of smoking on plasma vitamin C might be in part due to the inhibitory effect of smoke components on the synthesis of vitamin C in rats.

Leukocytes

In general, there is a relationship between serum or plasma and leukocyte vitamin C levels. Leukocyte levels are, however, believed to reflect tissue stores of the vitamin better than serum or plasma and are less affected by change in recent dietary intake. Unfortunately, the determination of vitamin C in leukocytes is technically difficult, making the procedure impractical for routine use. Leukocyte vitamin C concentrations of smokers are compiled from several papers (table 4).

Calder et al. [17] first demonstrated the lower vitamin C concentrations of leukocytes in smokers. Brook and Grimshaw [14] found that, as with plasma values, male smokers had significantly lower leukocyte concentrations than female smokers. Burr et al. [16] examined the vitamin C concentrations of leukocytes in elderly subjects aged 75 years and over. They found that vitamin C concentrations were lower in smokers than in nonsmokers, and that smokers consuming 15 cigarettes or more daily were lower in vitamin C concentration than smokers consuming less than 15 cigarettes daily. Albanese et al. [1] measured the leukocyte vitamin C concentrations by using an improved method, and confirmed the lower vitamin C concentrations in smokers.

Kevany et al. [57] studied leukocyte vitamin C concentrations and serum cholesterol levels in relation to smoking habits in middle aged males. Vitamin C concentrations in smokers were significantly lower than in nonsmokers, and significant correlations between vitamin C concentrations and serum cholesterol levels were found in smokers. McClean et al. [64] studied leukocyte vitamin C concentrations of smokers in relation to age, and found that the leukocyte vitamin C concentrations of smokers did not change significantly with increasing age, while the concentrations of nonsmokers appeared to decline with advancing age from 17–29 to 60–69 years. Schorah et al. [78] measured the leukocyte vitamin C concentrations in over a thousand pregnant women, and found that smokers had significantly lower vitamin C concentrations than nonsmokers.

Urine

Goyanna [33] reported that smoking more than 20 cigarettes daily entirely stopped the excretion of vitamin C in the urine. This is rather exaggerated, but there are several reports showing decreased excretion of vitamin C in the urine of smokers.

Pelletier [71–74] found that urinary excretion of vitamin C by smokers was about 60% of that by nonsmokers. However, no difference was found after saturation for 6 days with 2.2 g vitamin C. We measured the urinary vitamin C levels of smokers over a 4-year period, and found that the levels of smokers were about 70% those of nonsmokers [67, 68].

Others

The milk of smoking women contained much less vitamin C than that of women who did not smoke [95]. Placental tissue as well as maternal and umbilical cord plasma from women who smoked more than 5 cigarettes daily during pregnancy contained less than 50% of vitamin C than women

Table 5. Placental tissue, cord blood plasma and maternal plasma vitamin C levels of pregnant smokers and nonsmokers (adapted from Norkus et al. [69])

Group	n	Vitamin C, mg/dl		
		placental tissue	cord plasma	maternal plasma
Smokers	14	10.1 ± 4.8	0.61 ± 0.56	0.40 ± 0.32
Nonsmokers	12	20.9 ± 4.9	1.68 ± 0.58	0.89 ± 0.27

who did not smoke during pregnancy [69] (table 5). These data suggest that smoking during pregnancy places both mother and newborn at risk of low vitamin C status and its health consequences. The effects of experimentally produced cigarette smoke on tissue vitamin C concentrations have been studied.

Frogs and mice were exposed to cigarette smoke for 20 min daily. The animals showed a lowered vitamin C concentration of the adrenal glands, spleen, heart muscle, and lungs [94]. Guinea pigs were exposed to cigarette smoke for 20 min daily up to 29 days. The vitamin C concentration of the adrenal glands was significantly lower in animals receiving smoke than in control animals not exposed to smoke, but there was no significant difference in the vitamin C concentration of the testis and spleen in the two groups [27, 44]. Exposure of rats to cigarette smoke for 3–7 days caused a significant alteration in the vitamin C concentration of the lung [20].

On the other hand, long-term smoke exposure (16–24 weeks) did not significantly alter the concentrations of vitamin C in the lung, kidney and liver of rats and guinea pigs [21]. The lack of smoking effect on vitamin C status in rats could be partly attributable to the ability of rats to synthesize vitamin C. In the case of guinea pigs that cannot synthesize vitamin C, in addition to a metabolic adaptation following long-term smoke exposure, the abundant source of vitamin C (1,000 ppm) in the experimental diet used was probably the major factor responsible for the lack of smoking effect.

In both humans and hamsters, the vitamin C concentrations were significantly greater for alveolar macrophages from smokers compared with nonsmokers of the same species [66]. This might reflect protective utilization of vitamin C under conditions of increased oxidant stress, as mentioned later.

Acute Effect of Smoking on Vitamin C Level

In 1952 McCormick [65] stated that the smoking of one cigarette destroyed approximately 25 mg of vitamin C in the body, but lacked substantiating evidence. Calder et al. [17] were unable to find any decrease in plasma vitamin C in smokers and nonsmokers who smoked one cigarette every half hour over a period of six hours and in smokers who smoked 19–25 cigarettes in 6 h. Yeung [96] examined the vitamin C levels of plasma and leukocytes in young healthy adult females who smoked two cigarettes consecutively. There was little change in the levels between the presmoking and 20-min postsmoking samples. This result suggests that smoking has little or no acute effect on vitamin C levels of blood and leukocytes.

Sulochana and Arunagiri [89] investigated the acute effect of smoking on the excretion of vitamin C in urine in healthy smokers aged 20–30 years. The smokers, after collecting baseline urine, smoked 5 cigarettes within 1 h and the following 1 h urine was collected. In all the subjects, the urine collection showed higher vitamin C content, suggesting increased loss of vitamin C from the body.

Bourquin and Musmanno [12] studied the subacute effect of smoking on whole blood vitamin C levels. Male and female smokers smoked 10–40 cigarettes daily for 4 days. The blood level of vitamin C appeared to be lowered by 4-day smoking. However, this preliminary report used only three subjects and so is not conclusive.

In brief, the effect of smoking on vitamin C status is not acute, but long-term.

Dietary Intake of Vitamin C in Smokers

Information has been obtained about consumption of fruit and vegetables which are important sources of vitamin C, and dietary intake of vitamin C.

In a study by Calder et al. [17] there was no evidence from an approximate assessment of food habits that the difference in blood vitamin C levels was due to less intake of vitamin C among smokers. Pelletier [72] demonstrated that the lower blood vitamin C levels of smokers were not caused by lower vitamin C intakes.

Burr et al. [16] found in the elderly aged 75 years and over that smokers tended to eat fruit less often than nonsmokers, but the differences did not seem to account for the lower vitamin C levels in smokers. Data from the Nutrition Canada National Survey indicated a tendency for smokers to consume less vitamin C. However, the serum vitamin C levels for smokers were still lower than for nonsmokers when the effect of vitamin C intake was taken into account [73, 74]. Ritzel and Bruppacher [76] noted that citrus consumption appeared to be similar in smokers and nonsmokers. Keith and Driskell [53] found that vitamin C intakes for smokers and nonsmokers were not significantly different.

We also showed that there were no significant differences in fruit and vegetable consumption between smokers and nonsmokers, although smokers tended to have poorer consumption of fruit and vegetable. This suggests that dietary vitamin C intake for the two groups does not differ greatly [68].

Fehily et al. [28] studied associations between smoking habits and dietary vitamin intake by a 7-day weighed dietary survey, and found that smokers tended to have a lower intake of vitamin C. Keith and Mossholder [56] measured dietary intake of vitamin C from 24-hour food recalls, and found in adolescent females that smokers had lower dietary intake. However, when plasma vitamin C values were adjusted for vitamin C intake, smokers still exhibited significant lower plasma vitamin C levels. Norkus et al. [69] demonstrated that both smokers and nonsmokers had similar daily vitamin C intakes.

Smith and Hodges [81], using data from the second National Health and Nutrition Examination Survey of the United States, examined dietary intake and serum levels of vitamin C. Smokers had a lower mean daily intake (53 mg) of vitamin C than nonsmokers (65 mg) and a higher percentage (41 % compared to 31 %) consuming less than 70 % of the Recommended Dietary Allowance. However, median and mean serum vitamin C levels for smokers were consistently lower than for nonsmokers by approximately 0.2 mg/dl, when smokers and nonsmokers with similar dietary intakes of vitamin C were ranked by serum vitamin C levels. More recently, analyzing data from the same national survey, Schectman et al. [77] demonstrated that the inverse association between smoking and serum vitamin C levels occurred independently of dietary vitamin C intake, although smokers had decreased vitamin C intake compared to nonsmokers. In a study by Basu et al. [7], there was no apparent difference in vitamin C intake among smokers or nonsmokers, as estimated by dietary history of breakfast intake.

In conclusion, there are no significant differences in dietary intake of vitamin C between smokers and nonsmokers, although smokers tend to have poorer consumption of fruit and vegetable. Relatively lower intake of vitamin C by smokers may be partly responsible for the lower vitamin C status, but it is unlikely to explain more than a small part of the effects of smoking on vitamin C status.

Causes of Lowering Vitamin C Level by Smoking

Nicotine, a major toxicant of cigarette smoke, was found to accelerate the oxidation of vitamin C in vitro [12, 75]. Also in vitro cigarette smoke destroyed vitamin C in solution, but the vitamin was not destroyed by smoke from cigarette paper or by an equivalent amount of air drawn through the solution [17].

Pelletier [71, 72] observed that after saturation with vitamin C, the blood levels of both smokers and nonsmokers decreased at about the same rate when fed a diet low in vitamin C, and considered that there was no impairment in the smokers' metabolism of vitamin C. Instead, they suggested less efficient absorption of vitamin C by smokers. Keith and Pelletier [52] studied the in vivo effect of nicotine on vitamin C absorption and metabolism in guinea pigs. At 1 h after a dose of [1-^{14}C] ascorbic acid, there was more radioactivity and ascorbic acid remaining in the gastrointestinal tract contents of nicotine-treated animals than in control animals. There were lower concentrations of [1-^{14}C] ascorbic acid in adrenals, brain, kidneys and liver of nicotine-treated animals 1 h after the dose, but higher concentrations 3, 6 and 12 h after the dose when compared to control animals. They suggested that nicotine caused a delay in the absorption of the vitamin.

Kallner et al. [51] followed the time course of radioactivity in plasma and urine after oral administration of a single dose of [1-^{14}C] ascorbic acid in healthy male smokers consuming over 20 cigarettes per day. The study was carried out under steady-state conditions with regard to plasma vitamin C levels at intakes of 30 to 180 mg/day. The data were evaluated by pharmacokinetic principles, comparing them to data from healthy male nonsmokers [50]. Smokers had a higher metabolic turnover than nonsmokers at corresponding total turnover of vitamin C. In nonsmokers, metabolic turnover leveled off at about 40–50 mg/day, whereas in the smokers it leveled off at about 70–90 mg/day. This indicates that smokers have a significantly higher metabolic turnover than that found for nonsmokers; the turnover rate of smokers is increased by about 40% compared to that of nonsmokers. It is also noteworthy that the ratio between the amount of [1-^{14}C] ascorbic acid and total radioactivity in urine was less in smokers than in nonsmokers. This would indicate a higher degree of metabolism in smokers.

In addition, smokers had a somewhat impaired intestinal absorption of vitamin C (76.4 ± 9.3%, n = 6) when compared to nonsmokers (83.5 ± 3.3%, n = 7) [50, 51] as suggested by Pelletier [71, 72] and Keith and Pelletier [52].

Smokers are generally also drinkers. However, there is some evidence that moderate drinking (less than 50 g of pure alcohol) does not affect vitamin C status [11, 68].

Evidently the causes for lowered vitamin C levels in smokers could be mainly due to the higher metabolic turnover of vitamin C and partly to impaired intestinal absorption of the vitamin. In addition, inadequate dietary vitamin C intake, increased vitamin C utilization by the tissue, increased vitamin C excretion and direct oxidation of vitamin C by oxidants in smoke might also be factors.

A possible explanation why pipe and cigar smokers were found to have blood vitamin C levels comparable to those of nonsmokers might be because they did not inhale the smoke and consequently absorbed less nicotine and other toxicants [74].

Requirement of Vitamin C

Requirement in Man

Requirements of vitamin C in man have been debated almost since *L*-ascorbic acid was first recognized as the antiscorbutic vitamin in the 1930s. Several reviews have been published dealing with the requirement of vitamin C in man [30, 31, 41, 42, 49, 61, 70].

A daily intake of approximately 10 mg of vitamin C has been observed to be enough to prevent scurvy.

The present Recommended Dietary Allowances (RDAs) of vitamin C for adult men, 30–60 mg/day in most Western countries, may be regarded as the amount to be taken from the food, but it may not reflect the optimal amount for the best of health and for optimal regulation of all the biochemical functions of the vitamin. The recommendations have been derived mostly from blood levels and urinary excretion of vitamin C, and from studies on the intake of vitamin C to prevent the appearance of vitamin C deficiency symptoms. One reason why there are differences in the RDAs in different countries is that they are bureaucratic rather than scientific.

Another approach has been the saturation of body vitamin C. The RDA does not ensure maximum steady-state levels in the blood and even less so in the tissues. If maintaining a maximum body vitamin C pool is a desirable goal to optimal health, this can be achieved in men with an average daily vitamin C intake of 100 mg or more. Garry et al. [29] in healthy elderly males estimated this value to be 150 mg and Jacob et al. [47] in young adult males, 138 mg. However, some experimental findings as well as theoretical considerations suggest that the body vitamin C pool may be further enlarged by increasing the intake [5, 31, 32].

The requirement of vitamin C is also based on measurements of vitamin C catabolism, which generate estimates of daily vitamin C turnover. Kallner et al. [5], on the basis of a pharmacokinetic study on [1-^{14}C] ascorbic acid turnover in healthy men, estimated this value to be about 100 mg/day. Either way, what has remained unclear is what intake of vitamin C leads to the best of health.

Requirement in Smokers

Studies on smoking and vitamin C mentioned above suggest that smokers have different vitamin C requirements compared to nonsmokers.

Kallner et al. [51], using the same experimental protocol applied in the previous study for nonsmokers [50], estimated that smokers would need an additional vitamin C intake of at least 40 mg/day (a daily intake of at least 140 mg compared to about 100 mg for nonsmokers [50]). Based on their finding that the median serum vitamin C level of smokers was approximately 0.2 mg/dl less than that of nonsmokers for the same dietary vitamin C intake, Smith and Hodges [81] estimated that smokers would need to consume an additional 59 mg (95% confidence interval of 52–68 mg) daily intake. This compares to 65 mg (53–79 mg) per day when analysis is based on mean values to attain serum vitamin C levels comparable to nonsmokers. Schectman et al. [77] estimated that smokers would need an additional vitamin C intake of approximately 130 mg, by the regression curves in their figure (linear expression curves for smokers and nonsmokers showing relation between dietary vitamin C intake and serum levels).

In addition, based on their findings, other authors [30, 42, 67, 68, 73, 74] have suggested that smokers might have increased needs for vitamin C to compensate for the lowered levels. Perhaps smokers may have a vitamin C requirement approximately 2 times greater than that of nonsmokers.

In France, RDA of vitamin C for smokers is 120 mg compared to 80 mg for nonsmokers. Based on the available studies, a different RDA should be considered for smokers and nonsmokers in all the countries of the world, as well.

It appears that, among smokers, intake of vitamin C should be increased either through diet or supplementation. In addition, if smokers wish to saturate their body with vitamin C, much more vitamin C intake would be required. For example, in a preliminary study by Pelletier [71] smokers required 630 mg more vitamin C to reach saturation.

Amelioration of Vitamin C Status with Supplementation

At high dietary intakes of vitamin C (> 200 mg/day) serum vitamin C levels for nonsmokers approached a plateau of 1.3 mg/dl, while in smokers serum vitamin C levels approached a lower plateau (1.1 mg/dl), but not until dietary intake was 300 mg/day [80, 81]. There are several reports indicating elevated levels of vitamin C in the blood and urine by vitamin C supplement.

Pelletier [71–74] evaluated vitamin C blood levels before and after saturation for 5 days with 2.2 g vitamin C. The smokers initially had lower vitamin C blood levels and lower urinary excretion than the nonsmokers, but no differences were found after the saturation period, except at higher levels. Bailey et al. [4] showed that in smokers the plasma level rose after a supplementation for 5 days with 2 g. Ritzel and Bruppacher [76] noted that male smokers taking vitamin C supplement had sufficient plasma levels. Hoefel [38] observed that vitamin C supplementation for 10 days with 1 g raised the blood vitamin C levels in smokers, and also that the values tended to return to their pretrial level once the supplementary intake had ceased.

Serum vitamin C levels were analyzed in smokers and nonsmokers in relation to supplemental intake of vitamin C, using data from the second National Health and Nutrition Examination Survey of the United States. Among smokers, 21.9% took some supplement containing vitamin C. In the supplemented smokers the risk of severe or marginal vitamin C deficiency was decreased 4.5-fold compared to unsupplemented smokers, and was also lower compared to unsupplemented nonsmokers [77, 81].

More recently, we found over three years that plasma vitamin C levels for smokers taking vitamin C supplement (500–2,000 mg/day) were significantly higher than those for smokers not taking supplement (table 6). The

Table 6. Plasma vitamin C levels of smokers taking and not taking vitamin C supplement (Murata et al. [68])

Year	Plasma vitamin C, mg/dl				p
	taking	n	not taking	n	
1983	0.77 ± 0.27	25	0.49 ± 0.15	51	< 0.001
1984	0.89 ± 0.31	27	0.56 ± 0.13	48	< 0.001
1985	0.84 ± 0.18	19	0.56 ± 0.15	56	< 0.001

values of smokers taking supplement were apparently the same as those of nonsmokers who were not taking vitamin C supplement, but were lower than of nonsmokers who were taking vitamin C supplement. Also, urinary vitamin C levels of smokers taking vitamin C supplement were significantly higher than those of smokers not taking supplement [68].

These studies indicate that an adequate vitamin C supplement in addition to proper diet would assure satisfactory blood and urinary levels of vitamin C.

Effects of Supplementary Vitamin C in Smokers

Cigarette smoke contains a large number of toxic agents, and these are responsible for adverse effects of smoking, while vitamin C is a well known detoxifying agent and can scavenge toxic free radicals which are also generated by smoking. In addition, vitamin C has several pharmacologic functions when taken far in excess of physiological requirements (500 mg or more).

Protection against Toxicity of Smoke Components

Concerning protection against the toxicity of acetaldehyde and other toxicants in cigarette smoke, Sprince et al. [82–88] published a series of papers pointing increasingly to consideration of vitamin C as a possible protectant.

Acetaldehyde is quantitatively the highest ciliatoxic agent found in cigarette smoke and has been believed to be related to premalignancy of the tracheobronchial tree. Protection against acetaldehyde toxicity was

Table 7. Protection by vitamin C against acetaldehyde, acrolein, formaldehyde and cyanide toxicity in rat

Toxicant	Vitamin C mg/kg	Anesthesia in 3–10 min, %	Survivors after 72 h, %	Reference
Acetaldehyde	0	96	10	86
	176	78	41	
	352	45	75	
	528	14	90	
Acrolein	0	–	5	84
	528	–	95	
Formaldehyde	0	–	5	84
	528	–	55	
Cyanide	0	–	20	85
	528	–	100	

studied in rats by oral intubation of vitamin C 30–45 min prior to oral intubation of a standardized oral lethal dose of acetaldehyde. Animals were monitored for anesthesia and lethality for 72 h. Vitamin C showed moderate protection against anesthesia and marked protection against lethality [83, 86] (table 7). A combination of vitamin C with cysteine and thiamine gave virtually complete protection by either oral or intraperitoneal administration [86, 88]. Also, protection of rats against lethal synergy of acetaldehyde with nicotine was demonstrated by oral pretreatment with vitamin C [87].

Acrolein and formaldehyde, as well as acetaldehyde, are currently regarded as important toxicants in cigarette smoke. Protection of rats against acrolein and formaldehyde toxicity was also demonstrated [84]. Sprince et al. [86] have suggested that vitamin C could act directly against these toxicants per se or indirectly against the excessive release of catecholamines and corticosteroids induced by such toxicants.

Cyanide has been regarded as a respiratory tract toxicant of cigarette smoke. Also, protection against cyanide lethality in rats by vitamin C and dehydroascorbic acid, an oxidized form of vitamin C, was shown by Sprince et al. [85]. In this study, at lower doses dehydroascorbic acid was the better protectant, although pretreatment of rats with high oral doses of vitamin C or dehydroascorbic acid gave similar good protection.

In short, Sprince et al. have found that vitamin C provides protection against the toxicity of acetaldehyde, acrolein and formaldehyde and possibly against other toxicants such as nicotine, carbon monoxide, N-nitroso compounds, cadmium and polynuclear hydrocarbons including benz(a)pyrene in cigarette smoke. Further research along these lines is warranted.

These studies suggest that vitamin C may have a protective effect in some aspect of smoke inhalation toxicity.

Lung Protection

The lung is the primary organ at risk from the effects of inhaled cigarette smoke, and smoking has been implicated as a contributing factor to the causation and exacerbation of various respiratory diseases.

There is some evidence to suggest that smoking leads to chronic phagocyte activation and enhanced oxidant generation in the lungs. Smoking causes a peripheral leukocytosis and the lungs of smokers are reported to contain up to 3 times more macrophages and neutrophils than those of nonsmokers. It is also reported that alveolar macrophages and peripheral blood polymorphonuclear leukocytes from smokers generate increased amounts of reactive oxidants during exposure to stimulants of membrane-associated oxidative metabolism in vitro. Increased numbers of hyperactive macrophages and polymorphonuclear leukocytes in the lungs of smokers with consequent chronic generation of phagocyte-derived oxidants may be involved in the etiology of smoking-related lung diseases [2, 3, 92, 93].

One of the principal biochemical reactions of vitamin C is to scavenge toxic free radicals, including reactive oxidants. Theron and Anderson [92] observed that vitamin C scavenged phagocyte-derived reactive oxidants and protected the elastase inhibitory capacity of human α-1-protease inhibitor from polymorphonuclear leukocyte-mediated oxidative inactivation in vitro. More recently, they investigated the effects of vitamin C (3 g daily for 14 days) on the functional activity of serum α-1-protease inhibitor and oxidant release by blood phagocytes from smokers in a placebo-controlled, double-blind, crossover trial. The elastase inhibitory capacity of serum α-1-protease inhibitor remained unchanged, but the early-occurring extracellular luminol-enhanced chemiluminescence responses of leukoattractant-activated blood declined significantly during supplementation of vitamin C [93]. The results suggest that vitamin C may scavenge toxic oxidants released by hyperactivated phagocytes from smokers and consequently in the protection of serum α-1-protease inhibitor in the lungs [93].

Anderson et al. [2, 3] compared the bimodal pattern of N-formyl-methionyl-leucyl-phenylalanine-activated luminol-enhanced chemiluminescence with distinct early extracellular and late intracellular oxidative responses in polymorphonuclear leukocytes from smokers and nonsmokers, and found that vitamin C neutralized the activity of polymorphonuclear leukocyte-derived reactive oxidants. The results suggest that vitamin C in the lungs of smokers may regulate phagocyte-mediated oxygen toxicity. The alveolar macrophages from smokers' lungs have twice the concentration of vitamin C as those from nonsmokers [66]. This might reflect protective utilization of vitamin C under conditions of increased oxidants.

Leuchtenberger and Leuchtenberger [58] demonstrated that, in hamster lung cultures, vitamin C protected against or reversed abnormal growth and malignant transformation occurring after repeated exposure to cigarette smoke. The increase of lysosomes after vitamin C also pointed to the possible importance of lysosomes in protecting the culture against the enhancement of carcinogenesis by smoke. Later on, they reported that, in human lung cultures, when grown in the presence of vitamin C, lung cultures exposed to cigarette smoke showed a stimulation of growth and a significant decrease in mitotic abnormalities. Smoke-exposed human breast cancer cultures (SK-Br-3), when grown in the presence of vitamin C, also showed acceleration of growth of epithelial cells, significant reduction in mitotic abnormalities, and occurrence of pseudoglandular structures, indicating differentiation [59].

Harada et al. [37] made quantitative studies of biological responses in hamsters exposed to cigarette smoke and of effects of vitamin C supplement on smoke inhalation toxicity. The vitamin C-supplemented animals showed slightly improved body weight gain and food efficiency, significantly lower incidences of rhinitis, focal bronchial epithelial hyperplasia and bronchiolar adenomatoid lesion, and depressed alveolar macrophage mobilization as compared with those in the unsupplemented animals at the same dose of cigarettes.

In summary, vitamin C may protect smokers against some of the damage caused by oxidants present in and generated by the inhaled smoke.

With regard to lung function tests, some parameters appeared to be somewhat impaired in smokers as compared to nonsmokers [53]. However, in smoking healthy men, low dose vitamin C supplement (300 mg daily for 3 weeks) appeared to have little effect on the lung function test

[54]. Also, a single dose of vitamin C (400 or 1,200 mg) had little effect on the lung function parameters of male and female smokers who smoked two cigarettes within ten minutes [55].

Other Effects

Treadmill workload was significantly smaller for smokers, while no differences between smokers and nonsmokers were seen for oxygen consumption, minute ventilation, resting and post-exercise blood pressure and post-exercise blood lactic acid levels [53]. Bailey et al. [4] reported that short-term supplement of vitamin C (2 g daily for 5 days) to young smokers had little effect on respiratory adjustment and oxygen utilization before, during, and after treadmill exercise. Also, low dose vitamin C supplement (300 mg daily for 3 weeks) appeared to have little effect on physical performance (treadmill) of smokers [54]. Keith and Mossholder [55] studied the effect of a single dose of vitamin C (400 or 1,200 mg) on the acute physiological changes following the smoking of two cigarettes. 400 mg vitamin C resulted in a significant increase in systolic blood pressure following smoking, and 1,200 mg vitamin C resulted in a significant increase in diastolic blood response effect of vitamin C on the acute cardiovascular alterations caused by smoking.

There is much evidence that vitamin C may be of value in blocking the in vivo formation of N-nitroso compounds, which are known to be highly carcinogenic. Hoffmann and Brunnemann [40] documented the N-nitrosation potential of cigarette smoke which could lead to increased endogenous formation of N-nitrosamines in smokers, by measuring urinary excretion of N-nitrosoproline. This investigation also showed that a single dose of vitamin C (1 g) could inhibit the endogenous formation of N-nitrosoproline in smokers (table 8).

Table 8. Endogenous formation of N-nitrosoproline in smokers and nonsmokers (adapted from Hoffmann and Brunnemann [40])

Protocol	N-nitrosoproline, µg/24-h urine			
	smokers	n	nonsmokers	n
Controlled diet	5.9 ± 4.4	13	3.6 ± 2.1	13
Diet + proline	11.8	14	3.6	14
Diet + proline + vitamin C	4.6	13	4.7	13
Diet + vitamin C	6.0	8	4.0	9

Hamsters receiving subcutaneous injections of diethylnitrosamine were subjected to cigarette smoke inhalation and fed a diet with or without 1% vitamin C supplement for 58 weeks. The incidence of nasal cavity tumors and oral leukoplakic lesions was significantly lower in vitamin C-supplemented hamsters than in unsupplemented hamsters, while the development of tracheal and laryngeal tumors, and the induction of costochondral hyperplasia appeared to be accelerated by vitamin C [35, 36]. The situation is complex and requires further study, but the reported adverse effects may be insignificant in humans who take vitamin C supplement up to a few grams per day.

Conclusions

Impaired vitamin C status of cigarette smokers has been well documented in numerous publications since the 1940s. The effect of smoking on vitamin C status is not acute but long-term. The causes of the lowered vitamin C status in smokers could be due not to poorer dietary intake of vitamin C, but rather to the higher metabolic turnover of vitamin C and in part the impaired intestinal absorption of vitamin C.

Smokers have an increased risk of marginal vitamin C deficiency and its health consequences. They have a special requirement for vitamin C to compensate for their lowered levels. It is suggested that smokers may have a vitamin C requirement of approximately 2 times greater than that of nonsmokers. On the basis of the available studies, a different Recommended Dietary Allowance should be considered for smokers and nonsmokers, as in France.

Although complete cessation of smoking is the best solution, low-dose supplementation of vitamin C should be considered by those who cannot or do not want to give up smoking. The results of most studies support arguments in favor of vitamin C supplements in smokers, although the benefits of high-dose supplementation have not been assessed.

References

1   Albanese, A.A.; Wein, E.H.; Mata, L.A.: An improved method for determination of leucocyte and plasma ascorbic acid of man with applications to studies on nutritional needs and effects of cigarette smoking. Nutr. Rep. int *12:* 271–289 (1975).
2   Anderson, R.; Theron, A.J.; Ras, G.J.: Regulation by the antioxidants ascorbate,

cysteine, and dapsone of the increased extracellular and intracellular generation of reactive oxidants by activated phagocytes from cigarette smokers. Am. Rev. resp. Dis. *135:* 1027–1032 (1987).

3 Anderson, R.; Theron, A.J.; Ras, G.J.: Ascorbic acid neutralizes reactive oxidants released by hyperactive phagocytes from cigarette smokers. Lung *166:* 149–159 (1988).

4 Bailey, D.A.; Carron, A.V.; Teece, R.G.; Wehner, H.J.: Vitamin C supplementation related to physiological response to exercise in smoking and nonsmoking subjects. Am. J. clin. Nutr. *23:* 905–912 (1970).

5 Baker, E.M.; Saari, J.C.; Tolbert, B.M.: Ascorbic acid metabolism in man. Am. J. clin. Nutr. *19:* 371–378 (1966).

6 Basu, T.K.; Schorah, C.J.: Vitamin C in health and disease (Croom Helm, London 1982).

7 Basu, J.; Vermund, S.H.; Mikhail, M.; Palan, P.R.; Romney, S.L.: Plasma reduced and total ascorbic acid in healthy women: effects of smoking and oral contraception. Contraception *39:* 85–93 (1989).

8 Bazzarre, T.L.: Effect of vitamin C supplementation among male smokers. Nutr. Rep. int. *33:* 711–720 (1986).

9 Bendich, A.; Machlin, L.J.; Scandurra, O.; Burton, G.W.; Wayner, D.D.M.: The antioxidant role of vitamin C. Adv. Free Radical biol. Med. *2:* 419–444 (1986).

10 Biersner, R.J.; Gilman, S.C.; Thornton, R.D.: Relationship of plasma vitamin C to the health and performance of submariners. J. appl. Nutr. *34:* 29–37 (1982).

11 Bonjour, J.P.: Vitamins and alcoholism. I. Ascorbic acid. Int. J. Vitam. Nutr. Res. *49:* 434–441 (1979).

12 Bourquin, A.; Musmanno, E.: Preliminary report on the effect of smoking on the ascorbic acid content of whole blood. Am. J. dig. Dis. *20:* 75–77 (1953).

13 Brin, M.: Marginal vitamin C deficiency and human health; in Counsell, Hornig, Vitamin C; ascorbic acid, pp. 359–376 (Applied Science Publishers, London 1981).

14 Brook, M.; Grimshaw, J.J.: Vitamin C concentration of plasma and leukocytes as related to smoking habit, age, and sex of humans. Am. J. clin. Nutr. *21:* 1254–1258 (1968).

15 Burns, J.J.; Rivers, J.M.; Machlin, L.J. (eds.): Third Conference on Vitamin C. Ann. N.Y. Acad. Sci., vol. 498 (1987).

16 Burr, M.L.; Elwood, P.C.; Hole, D.J.; Hurley, R.J.; Hughes, R.E.: Plasma and leukocyte ascorbic acid levels in the elderly. Am. J. clin. Nutr. *27:* 144–151 (1974).

17 Calder, J.H.; Curtis, R.C.; Fore, H.: Comparison of vitamin C in plasma and leucocytes of smokers and nonsmokers. Lancet *i:* 556 (1963).

18 Chatterjee, I.B.: Ascorbic acid metabolism. Wld Rev. Nutr. Diet. *30:* 69–87 (1978).

19 Chen, L.H.; Chow, C.K.: Effect of cigarette smoking and dietary vitamin E on plasma level of vitamin C in rats. Nutr. Rep. Int. *22:* 301–309 (1980).

20 Chow, C.K.; Bridges, R.B.: Chronic cigarette smoking and plasma micronutrients. Fed. Proc. *43:* 861 (1984).

21 Chow, C.K.; Airriess, G.R.; Chen. L.-C.; Changchit, C.: Vitamin C levels in the tissues of cigarette-smoked guinea pigs and rats. Ann. N.Y. Acad. Sci. *498:* 467–469 (1987).

22 Chow, C.K.; Thacker, R.R.; Changchit, C.; Bridges, R.B.; Rehm, S.R.; Humble, J.; Turbek, J.: Lower levels of vitamin C and carotenes in plasma of cigarette smokers. J. Am. Col. Nutr. 5: 305–312 (1986).

23 Clemetson, C.A.B.: Vitamin C (CRC Press, Boca Raton 1989).

24 Counsell, J.N.; Hornig, D.H. (eds.): Vitamin C; ascorbic acid (Applied Science Publishers, London 1981).

25 Elwood, P.C.; Hughes, R.E.; Hurley, R.J.: Ascorbic acid and serum cholesterol. Lancet ii: 1197 (1970).

26 Englard, S.; Seifer, S.: The biochemical functions of ascorbic acid. Ann. Rev. Nutr. 6: 365–406 (1986).

27 Evans, J.R.; Hughes, R.E.; Jones, P.R.: Some effects of cigarette smoke on guinea-pigs. Proc. Nutr. Soc. 26: 36 (1967).

28 Fehily, A.M.; Phillips, K.M.; Yarnell, J.W.G.: Diet, smoking, social class, and body mass index in the caerphilly heart disease study. Am. J. clin. Nutr. 40: 827–833 (1984).

29 Garry, P.J.; Goodwin, J.S.; Hunt, W.C.; Gilbert, B.A.: Nutritional status in a healthy elderly population: vitamin C. Am. J. clin. Nutr. 36: 332–339 (1982).

30 Gerster, H.: Human vitamin C requirements. Z. ErnährWiss 26: 125–137 (1987).

31 Ginter, E.: Chronic marginal vitamin C deficiency: biochemistry and pathophysiology. Wld Rev. Nutr. Diet. 33: 104–141 (1979).

32 Ginter, E.: What is truly the maximum body pool size of ascorbic acid in man. Am. J. clin. Nutr. 33: 538–539 (1980).

33 Goyanna, C.: Tobacco and vitamin C. Brasil Med. 69: 173–177 (1955).

34 Hanck, A. (ed.): Vitamin C. New clinical applications in immunology, lipid metabolism and cancer. Int. J. Vitam. Nutr. Res., suppl. 23, pp. 1–294 (1982).

35 Harada, T.; Enomoto, A.; Kitazawa, T.; Maita, K.; Shirasu, Y.: Oral leukoplakia and costochondral hyperplasia induced by diethylnitrosamine in hamsters exposed to cigarette smoke with or without dietary vitamin C. Vet. Pathol. 24: 257–264 (1987).

36 Harada, T.; Kitazawa, T.; Maita, K.; Shirasu, Y.: Effects of vitamin C on tumor induction by diethylnitrosamine in the respiratory tract of hamsters exposed to cigarette smoke. Cancer Letters 25: 163–169 (1985).

37 Harada, T.; Maita, K.; Nakashima, N.; Odanaka, Y.; Shirasu, Y.: Quantitative studies of biological responses in hamsters exposed to tobacco smoke and effects of vitamin C supplement on smoke inhalation toxicity. Jap. J. vet. Sci. 45: 613–626 (1983).

38 Hoefel, O.S.: Plasma vitamin C levels in smokers. Int. J. Vitam. Nutr. Res., suppl. 16, pp. 127–137 (1977).

39 Hoefel, O.S.: Smoking; an important factor in vitamin C deficiency. Int. J. Vitam. Nutr. Res., suppl. 24, pp. 121–124 (1983).

40 Hoffmann, D.; Brunnemann, K.D.: Endogenous formation of N-nitrosoproline in cigarette smokers. Cancer Res. 43: 5570–5574 (1983).

41 Hornig, D.: Requirement of vitamin C in man. Trends Pharmacol. Sci. 3: 294–296 (1982).

42 Hornig, D.H.; Glatthaar, B.E.: Vitamin C and smoking; increased requirement of smokers. Int. J. Vitam. Nutr. Res., suppl. 27, pp. 139–155 (1985).

43 Huges, R.E.: Vitamin C; Some current problems (British Nutrition Foundation, London 1981).

44 Hughes, R.E.; Jones, P.R.; Nicholas, P.: Some effects of experimentally-produced cigarette smoke on the growth, vitamin C metabolism and organ weights of guinea-pigs. J. Pharm. Pharmac. *22:* 823–827 (1970).

45 Information Canada: Nutritional Canada interpretive standards, national survey. (Information Canada, Ottawa 1973).

46 Irwin, M.I.; Hutchins, B.K.: A conspectus of research on vitamin C requirements of man. J. Nutr. *106:* 823–879 (1976).

47 Jacob, R.A.; Skala, J.H.; Omaye, S.T.: Biochemical indices of human vitamin C status. Am. J. clin. Nutr. *46:* 818–826 (1987).

48 Jaffe, G.M.: Vitamin C; in Machlin, Handbook of vitamins, pp. 199–244 (Marcel Dekker, New York 1984).

49 Kallner, A.: Requirement for vitamin C based on metabolic studies. Ann. N.Y. Acad. Sci. *498:* 418–423 (1987).

50 Kallner, A.; Hartmann, D.; Hornig, D.: Steady-state turnover and body pool of ascorbic acid in man. Am. J. clin. Nutr. *32:* 530–539 (1979).

51 Kallner, A.B.; Hartmann, D.; Hornig, D.H.: On the requirements of ascorbic acid in man: steady-state turnover and body pool in smokers. Am. J. clin. Nutr. *34:* 1347–1355 (1981).

52 Keith, M.O.; Pelletier, O.: The effect of nicotine on ascorbic acid retention by guinea pigs. Can. J. Physiol. Pharmacol. *51:* 879–884 (1973).

53 Keith, R.E.; Driskell, J.A.: Effects of chronic cigarette smoking on vitamin C status, lung function, and resting and exercise cardiovascular metabolism in humans. Nutr. Rep. Int. *21:* 907–912 (1980).

54 Keith, R.E.; Driskell, J.A.: Lung function and treadmill performance of smoking and nonsmoking males receiving ascorbic acid supplements. Am. J. clin. Nutr. *36:* 840–845 (1982).

55 Keith, R.E.; Mossholder, S.B.: Vitamin C and acute physiological responses to cigarette smoking. Nutr. Res. *3:* 653–661 (1983).

56 Keith, R.E.; Mossholder, S.B.: Ascorbic acid status of smoking and nonsmoking adolescent females. Int. J. Vitam. Nutr. Res. *56:* 363–366 (1986).

57 Kevany, J.; Jessop, W.; Goldsmith, A.: The effect of smoking on ascorbic acid and serum cholesterol in adult males. Ir. J. med. Sci. *144:* 474–477 (1975).

58 Leuchtenberger, C.; Leuchtenberger, R.: Protection of hamster lung cultures by *L*-cysteine or vitamin C against carcinogenic effects of fresh smoke from tobacco or marihuana cigarettes. Br. J. exp. Path. *58:* 625–634 (1977).

59 Leuchtenberger, C.; Leuchtenberger, R.: The effects of naturally occurring metabolites (*L*-cysteine, vitamin C) on cultured human cells exposed to smoke of tobacco or marijuana cigarettes. Cytometry *5:* 396–402 (1984).

60 Levine, M.: New concepts in the biology and biochemistry of ascorbic acid. New Engl. J. Med. *314:* 892–902 (1986).

61 Levine, M.; Hartzell, W.: Ascorbic acid; the concept of optimum requirements. Ann. N.Y. Acad. Sci. *498:* 424–444 (1987).

62 Levine, M.; Morita, K.: Ascorbic acid in endocrine systems. Vitams Horm. *42:* 1–64 (1985).

63 Lewin, S.: Vitamin C; Its molecular biology and medical potential (Academic Press, New York 1976).

64 McClean, H.E.; Dodds, P.M.; Abernethy, M.H.; Stewart, A.W.; Beaven, D.W.: Vitamin C concentration in plasma and leucocytes of men related to age and smoking habit. N.Z. med. J. *83:* 226–229 (1976).

65 McCormick, W.J.: Ascorbic acid as a chemotherapeutic agent. Archs Pediat. *69:* 151–155 (1952).

66 McGowan, S.E.; Parenti, C.M.; Hoidal, J.R.; Niewoehner, D.E.: Ascorbic acid content and accumulation by alveolar macrophages from cigarette smokers and nonsmokers. J. Lab. clin. Med. *104:* 127–134 (1984).

67 Murata, A.; Morinaga, N.; Kato, F.; Harada, Y.: Plasma and urine vitamin C levels in male smokers at periodic health examinations. Vitamins (Kyoto) *58:* 61–69 (1984).

68 Murata, A.; Shiraishi, I.; Fukuzaki, K.; Kitahara, T.; Harada, Y.: Lower levels of vitamin C in plasma and urine of Japanese male smokers. Int. J. Vitam. Nutr. Res. *31:* 184–189 (1989).

69 Norkus, E.P.; Hsu, H.; Cehelsky, M.R.: Effect of cigarette smoking on the vitamin C status of pregnant women and their offspring. Ann. N.Y. Acad. Sci. *498:* 500–501 (1987).

70 Olson, J.A.; Hodges, R.E.: Recommended dietary intakes (RDA) of vitamin C in humans. Am. J. clin. Nutr. *45:* 693–703 (1987).

71 Pelletier, O.: Smoking and vitamin C levels in humans. Am. J. clin. Nutr. *21:* 1259–1267 (1968).

72 Pelletier, O.: Vitamin C status of cigarette smokers and nonsmokers. Am. J. clin. Nutr. *23:* 520–524 (1970).

73 Pelletier, O.: Vitamin C and cigarette smokers. Ann. N.Y. Acad. Sci. *258:* 156–168 (1975).

74 Pelletier, O.: Vitamin C and tobacco. Int. J. Vitam. Nutr. Res., suppl. 16, pp. 147–169 (1977).

75 Reif, G.: Versuche über das Verhalten von Nicotin und Coffein gegen Ascorbinsäure. Biochem. Z. *315:* 310–319 (1943).

76 Ritzel, G.; Bruppacher, R.; Vitamin C and tabacco. Int. J. Vitam. Nutr. Sci., suppl. 16, pp. 171–183 (1977).

77 Schectman, G.; Byrd, J.C.; Gruchow, H.W.: The influence of smoking on vitamin C status in adults. Am. J. publ. Hlth *79:* 158–162 (1989).

78 Schorah, C.J.; Zemroch, P.J.; Sheppard, S.; Smithells, R.W.: Leucocyte ascorbic acid and pregnancy. Br. J. Nutr. *39:* 139–149 (1978).

79 Seib, P.A.; Tolbert, B.M.: Ascorbic acid: chemistry, metabolism and uses (American Chemical Society, Washington 1982).

80 Smith, J.L.; Hodges, R.E.: Quantitative relationship between serum levels and dietary intake of vitamin C in smokers and nonsmokers. Fed. Proc. *44:* 778 (1985).

81 Smith, J.L.; Hodges, R.E.: Serum levels of vitamin C in relation to dietary and supplemental intake of vitamin C in smokers and nonsmokers. Ann. N.Y. Acad. Sci. *498:* 144–152 (1987).

82 Sprince, H.: Ascorbic acid, sulfur compounds, and anti-adrenergic agents as protectants against acetaldehyde toxicity: implications in alcoholism and smoking. Br. J. Alcohol Alcohol. *16:* 5–9 (1981).

83 Sprince, H.; Parker, C.M.; Smith, G.F.: Acetaldehyde, ascorbic acid, and catecholamine-regulating drugs: data and hypothesis in relation to alcoholism and smoking. Nutr. Rep. Int. *17:* 441–455 (1978).

84 Sprince, H.; Parker, C.M.; Smith, G.G.: Comparison of protection by *L*-ascorbic acid, *L*-cysteine, and adrenergic-blocking agents against acetaldehyde, acrolein, and formaldehyde toxicity: implications in smoking. Agent Action *9:* 407–414 (1979).

85 Sprince, H.; Smith, G.G.; Parker, C.M.; Rinehimer, D.A.: Protection against cyanide lethality in rats by *L*-ascorbic acid and dehydroascorbic acid. Nutr. Rep. Int. *25:* 463–470 (1982).

86 Sprince, H.; Parker, C.M.; Smith, G.G.: *L*-Ascorbic acid in alcoholism and smoking: protection against acetaldehyde toxicity as an experimental model. Int. J. Vitam. Nutr. Res., suppl. 16, pp. 185–211 (1977).

87 Sprince, H.; Parker, C.M.; Smith, G.G.: Lethal synergy of acetaldehyde with nicotine, caffeine, or dopamine in rats: protection by ascorbic acid, cysteine, and antiadrenergic agents. Nutr. Rep. Int. *23:* 43–54 (1981).

88 Sprince, H.; Parker, C.M.; Smith, G.G.; Gonzales, L.L.: Portective action of ascorbic acid and sulfur compounds against acetaldehyde toxicity: implications in alcoholism and smoking. Agent Action *5:* 164–173 (1975).

89 Sulochana, G.; Arunagiri, R.: Smoking and ascorbic acid excretion. Clinician *45:* 198–201 (1981).

90 Surgeon General: Smoking and health (US Public Health Service, Rockville 1979).

91 Surgeon General: The health consequences of smoking. (US Public Health Service, Rockville 1982).

92 Theron, A.; Anderson, R.: Investigation of the protective effects of the antioxidants ascorbate, cysteine, and dopsone on the phagocyte-mediated oxidative inactivation of human alpha-1-protease inhibitor in vitro. Am. Rev. resp. Dis. *132:* 1049–1054 (1985).

93 Theron, A.J.; Anderson, R.: Investigation of the effects of oral administration of ascorbate on the functional activity of serum alpha-1-protease inhibitor and oxidant release by blood phagocytes from cigarette smokers in a placebo-controlled, double-blind, crossover trial. Int. J. Vitam. Nutr. Res. *58:* 218–224 (1988).

94 Venulet, F.; Tabacco smoke and ascorbic acid. Endokrinologie *30:* 345–351 (1953); cited in Chem. Abstr. 48, 8354a (1954).

95 Venulet, F.; Danysz, A.: Effect of tobacco smoking on the vitamin C content of human milk. Pediatria pol *31:* 811–817 (1955); cited in Chem. Abstr. 51, 8977g (1957).

96 Yeung, D.L.: Relationship between cigarette smoking, oral contraceptives, and plasma vitamins A, E, C and plasma triglycerides and cholesterol. Am. J. clin. Nutr. *29:* 1216–1221 (1976).

Akira Murata, PhD, Department of Applied Biological Sciences, Saga University, Saga 840 (Japan)

Simopoulos AP (ed): Selected Vitamins, Minerals, and Functional Consequences of
Maternal Malnutrition. World Rev Nutr Diet. Basel, Karger, 1991, vol 64, pp 58–84

Retinoids in the Host Defense System

Manabu Yamamoto

Department of Food and Nutrition, Tachikawa College of Tokyo, Japan

Contents

I. Introduction

Vitamin A is essential for life maintenance. Besides its mode of action in vision, the knowledge about the roles of vitamin A and retinoids (its natural metabolites or synthetic analogs) pertaining to normal growth and cell differentiation, has markedly increased [11, 34, 36, 40, 70, 98, 108, 119, 124]. Thus, it has been proposed recently that retinoic acid may act as morphogen in the developing chick limb bud [99, 114]. Furthermore, it has been indicated that human retinoic acid receptor belongs to the family of nuclear receptors, which are similar to steroid hormones and thyroid hormones, and retinoids can control genomic expression of cells [15, 16, 21, 35, 51, 89]. Considering these critical roles of retinoids, it is not surprising that retinoids also have a critical role in disease prevention or immunity.

Actually recent studies have also concentrated on the beneficial action of retinoids in health and disease. From this point of view, these studies can be separated in the following two groups. In the first group of studies, many investigatons focused on the action of retinoids in infectious diseases or immunity [8, 9, 34, 39, 60, 106]. Resistance to infection is deeply affected by vitamin A and its deficiency is a threat to human health, especially in developing countries [25, 69, 106]. Thus, clinical trials supplying vitamin A have been conducted and its effects evaluated. On the other hand, recent progress in biological sciences, such as immunology and genetics, provides new methods and techniques in this area. Thus, present knowledge on the roles of retinoid action for the resistance to infection is also increased.

In another group of studies, investigators have studied the action of retinoids on neoplasms. The formation of cancer has been found to increase in vitamin-A-deficient animals [1, 11, 18, 33, 40, 82, 108]. Epidemiological studies indicate that the incidence of cancer is inversely correlated with daily intakes of vitamin A or carotene [1, 41, 42, 86, 98, 108]. Thus, pharmacologists are looking for a new retinoid which has a high cancer-preventive action and low toxicity. Their mode of action on tumor initiation and promotion has also been studied.

Thus, the interest in the action of retinoids now extends widely from nutritionists to researchers in the medical and biological sciences. From the nutritional point of view, these studies are separated into two groups, that is, studies on the physiological role of retinoids, such as its action on vision, and studies on the effect of excess retinoid or synthetic retinoid, such as its action as an adjuvant material.

The purpose of this article is to review the present knowledge about the roles of vitamin A and retinoids from an immunological aspect.

II. Retinoids and Disease Prevention

This section is to provide a perspective on the role of retinoids for disease prevention. More detailed reviews on the specific problem, such as retinoids and cancer, or carotenoid and cancer, will be found elsewhere [40, 53, 55, 56, 70, 108].

A. Resistance to Infection and Retinoids
1. Effect of Vitamin A Deficiency in Experimental Animals

Historically, in 1917, McCollum [58] had already noticed frequent development of infections in vitamin-A-deficient rats. Green and Mellanby [37, 38] reported that severe infection in vitamin-A-deficient rats was reversed by vitamin A administration. Thus, they called vitamin A an anti-infective agent. Then, accumulated evidence indicated that vitamin-A-deficient animals were susceptible to infections. Many investigations on the mechanisms of vitamin A have also been carried out in deficient animals. With respect to pathogenic agents, infections due to bacterial, protozoal, and viral origin were examined.

In 1968, Scrimshaw et al. [92] evaluated nearly fifty investigations on diseases, in which vitamin A deficiency resulted in greater frequency, severity, or fatality. They concluded in the review that no nutritional deficiency is more consistently synergistic with infectious disease than vitamin A deficiency, and one of the first recognized features of avitaminosis A was increased susceptibility to infection. Although many of these results had no pair fed controls, some investigators have suggested that vitamin-A-deficient animals are significantly more suspectible to infections than pair-fed controls [79, 92].

On the other hand, Bieri et al. [7] induced vitamin A deficiency in a germ-free state. They found that death usually occurred by the 8th week postweaning in conventionalized vitamin-A-deficient rats. In a germ-free state, although the growth ceased by the 8th week and symptoms of vitamin A deficiency appeared in vitamin-A-deficient rats, the plateau pattern for weight was maintained for a long period, as long as 40 weeks in some cases. This indicated that life can be maintained on extremely low levels of vitamin A, in the absence of infection or possible environmental stress

[91]. Considering these results, it seems that the essentiality of vitamin A in the maintenance of life may largely be mediated by its effect on resistance to infection.

2. Effect of Vitamin A Deficiency in Human Studies

In human studies, children and adults with xerophthalmia have been shown to be highly susceptible to natural infections, such as tuberculosis, measles and genito-urinary infections [4, 25, 44, 69, 92, 103, 106]. However, we have to keep in mind that human vitamin A deficiency is usually associated with other nutrient deficiencies.

Recently, Sommer et al. [102] have conducted a large epidemiological and clinical trial in Indonesia. In 1983, they reported that mild vitamin A deficiency was directly associated with at least 16% of all deaths in children from 1 to 6 years of age, when diagnosed as mild xerophthalmia. They further indicated that the incidences of diarrhea and respiratory disease were higher among xerophthalmic children and were independent of their nutritional status [103]. Sommer et al. [104] carried out a randomized community trial with 25,939 preschool children by supplying a capsule containing 200,000 IU vitamin A, in 1982. In the report, they suggested that supplements given to vitamin-A-deficient populations may decrease mortality by as much as 34%. Similar results were also reported in other developing countries [25, 69, 73, 111]. The risk of diseases, such as respiratory disease, diarrhea, and measles is increased with vitamin A deficiency. Since the risk of vitamin A deficiency also increases with these disorders, it is possible to establish a cycle between these disorders and vitamin A deficiency among children with these disorders [105].

3. Comments

As reviewed above, evidence indicates that vitamin A is essential for maintaining the resistance to infection. Epidemiological trials, on the other hand, have demonstrated that vitamin A deficiency still remains a threat to the health and lives of millions of children. On the prophylactic value of vitamin A administration, it is interesting that a randomized trial has revealed a higher risk of mortality among the trial children with no capsule, when compared with the non-trial controls [111]. To make these results clearer, double-blind study with placebo capsules is needed. However, present results suggest the importance of both vitamin A nutrition and nutritional education for children's health in the developing countries.

On the other hand, Cohen and Elin [17] reported that the administration of 3,000 IU vitamin A induced nonspecific resistance to infection in mice. This result suggests that supplemental doses of vitamin A may also have a beneficial effect on resistance to infection. Thus, among the multiple functions of vitamin A, its action on resistance to infection seems important in human health and disease prevention.

B. Cancer Prevention and Retinoids
1. Studies on Experimental Animals

Retinoids can act not only in resistance to infection, but also in cancer prevention. These compounds are assumed to modify the differentiation of preneoplastic cells or the immune response to neoplastic cells [28, 29, 40, 55, 56, 70, 107, 108]. Thus, extensive studies have been carried out on experimental animals, using thousands of retinoids [28, 29, 55, 56]. Tumor inhibition by retinoids has been indicated among tumors, which were induced chemically, spontaneously, or virally. The purpose of this section is to summarize the effects from a dietary point of view, that is, the effect of vitamin A deficiency and the effect of oral supplementation with retinoids.

In 1926, Fujimaki [33] reported the formation of gastric carcinomas in vitamin-A-deficient rats. Nettesheim et al. [81, 82] reported that, following the injection of a carcinogen, 42% of rats given 10 µg per week of retinyl acetate developed pre-cancerous changes in the lung. On the other hand, only 3% of the rats developed these changes when they received 10,000 µg of retinyl acetate once a week. Since the rats were maintained on a vitamin-A-deficient diet, it was suggested that mild vitamin A deficiency enhanced the susceptibility to carcinogens. Concerning cancer of the urinary bladder, Cohen et al. [18] found that rats fed a vitamin-A-deficient diet were more susceptible to bladder carcinogenesis than rats receiving physiological levels of vitamin A. Davies [20] reported that the rate of both appearance and regression of papilloma, induced by single topical application of a carcinogen to the skin, were increased in mice fed a diet containing 100 IU/g diet, when compared with animals fed a vitamin-A-deficient diet. Similar findings in vitamin-A-depleted rats were also reported in colon cancer with aflatoxin B_1 [83]. Thus, vitamin A deficiency appears to increase the incidence and growth of cancers in experimental animals.

The effects of excess vitamin A or retinoid administration on induced cancers have also been extensively studied. The chemopreventive effects of vitamin A, retinoic acid, and its synthetic analogs were reported in cancers of skin, lung, breast, oral cavity, gastrointestinal tract and other sites [40, 55, 56, 108]. On the oral supplementation of vitamin A, Forni et al. [32] have recently reported the effect of prolonged supplementation with retinyl palmitate, ranging from 50 to 1,000 IU/day in mice. They found that the growth of 3 transplantable tumors was impaired when the challenge was performed on days 75 and 150.

These results indicate that vitamin A may act to modify the precancerous changes in normal tissue and to modify susceptibility to cancer induction. Furthermore, the administration of excess vitamin A or retinoids may also cause regression of induced tumors or delay the appearance of tumors in experimental animals.

Since then, a large number of retinoids has been screened as possible anti-cancer agents in experimental tumor models. Extensive studies have been examined involving cancer of skin, lung, mammary gland, gastrointestinal tract and others.

2. Retinoids and Incidence of Cancer in Humans

Epidemiological studies also suggest that there is an inverse relationship between the intake of vitamin A and cancer risk [41, 42, 86]. Most of these studies estimated dietary intake of vitamin A from the consumption of foods which had a high content of vitamin A or carotene. Thus, to interpret these data, it must be considered whether the compound of interest is retinol, retinoic acid or carotenoid. The conversion of β-carotene to retinol is estimated to be about one sixth, and excessive carotene may result in hypercarotenemia. Recently, it has been reported that some carotenoid, regardless of its vitamin A activity, can inhibit skin tumors which have been induced by carcinogens [53].

Presumably, the incidence of lung cancer is related to vitamin A or carotene consumption. Bjelke [10] reported that the consumption of vegetables and dairy products lowered the incidence of lung cancer, after adjusting for the effects of smoking. Hirayama [41, 42] conducted a comprehensive prospective study on a Japanese adult population of 265,118 from 1965. He found that the incidence of lung cancer was about half in persons with daily intakes of green-yellow vegetables. He further checked the incidence of stomach cancer and found similar results. Furthermore, daily intake of milk was also inversely correlated with the incidence of stomach cancer.

Studies on the relationship between the risk of cancer and serum retinol levels were also performed. Kark et al. [47, 48] measured serum retinol levels from 3,102 subjects, and found lower serum retinol levels in subjects who eventually developed cancer. Wald et al. [116] conducted a prospective study in a population of 16,000. They also found an association between low retinol levels and increased risk of cancer.

3. Comments

Retinoids can modify cell differentiation and immune competence. Thus, it seems reasonable to assume that retinoids can also act on preneoplastic cells by interfering with tumor induction by carcinogens. Retinoids can also act on the immune system to increase antitumor activity. The precise mechanism of these actions is still to be determined. However, these results, especially from the epidemiological studies described above may suggest the important role of retinoids or carotenoid in cancer prevention.

Many clinical studies on retinoids have also been conducted with premalignant skin lesions. However, in this review, these studies will only be discussed from an immunological aspect. More comprehensive reviews on these points can be found elsewhere [40, 55, 56, 70].

III. Morphological Changes of Lymphoid Organs in Vitamin A Deficiency

Systemic atrophy of lymphoid organs has been reported in vitamin-A-deficient animals [34, 54, 78, 80, 88, 122]. Among the lymphoid organs, in which atrophy was observed, were the thymus, spleen and lymph nodes. However, there are some difficulties in interpreting these data. First, in studies lacking pair-fed control animals, it is difficult to separate the effect of vitamin A per se from its effect on appetite [2]. Vitamin A deficiency increases susceptibility to infections as reviewed above and some of these infections may have passed unrecognized in the studies. Thus, differences in environmental stress and animal hygiene may change both the time course and results of experimental vitamin A deficiency.

A. Thymus

Krishnan et al. [54] were first to report marked thymus atrophy in vitamin-A-deficient rats. They observed significant reduction in thymus weight in vitamin-A-deficient rats, on the basis of body weight. Upon histological examination of these thymuses, they further found a depletion of lymphocytes from the cortex. Nauss et al. [78] also found a reduction in thymus weight in vitamin-A-deficient rats, but this was not significant on the basis of body weight. On the other hand, Chandra and Au [14] have indicated that no difference in thymus weight was found between vitamin-A-deficient rats and pair-fed controls. Similar results were also obtained in our study [120]. The supplemental effect of retinoids was examined in vitamin-A-deficient rats, using a plateau stage for weight. Although thymus weights increased in vitamin-A-supplemented rats, no significant difference in weight was seen between pair-fed controls and vitamin-A-deficient rats. Thus, atrophy of the thymus, due to vitamin A deficiency, appears to be mainly mediated by anorexia, which coexists with vitamin A deficiency [2]. Interestingly enough, in the studies which indicated the difference in thymus weight between vitamin-A-deficient animals and pair-fed controls, the average thymus weight was higher than that of the spleen in pair-fed controls.

B. Spleen

Although splenic atrophy was also found by Krishnan et al. [54] in vitamin-A-deficient rats, this has not been confirmed by other investigators. The weight of spleens from vitamin-A-deficient rats was significantly low when compared with that from vitamin-A-supplemented rats, but no difference in splenic weight was seen between the pair-fed control group and the vitamin-A-deficient group in our study (table 1). These same findings have also been reported by Nauss et al. [80]. They studied splenic changes during progressive stages of vitamin A deficiency in rats and no changes in relative splenic weight was found in their study. Thus, they concluded that the marked alterations of the thymus and spleen reported in rats with severe vitamin A deficiency could be partially accounted for by the association of anorexia. Probably, protein-energy malnutrition may be essential in causing atrophy in these organs, and vitamin A deficiency per se may not act to reduce these organ weights, when protein-energy malnutrition is not associated with vitamin A deficiency. Recently, Pasatiempo et al. [88] examined the effect of vitamin A deficiency using both sexes of Lewis rats. They found that relative spleen weight was reduced in vitamin-A-deficient rats, when deficiency was induced in male Lewis rats, but this was not seen in female rats. Splenomegaly has also been reported by Smith et al. [101] in severe vitamin A deficiency in mice. Since this splenomegaly was more predominant in SPF (specific pathogen-free) animals, they rather attributed this to the difference in species than the effect of infec-

Table 1. Effect of retinoid injection on spleen weight

Group	n	Spleen weight	
		mg	mg/100 g body weight
Experiment I			
A DEF	9	438 ± 54	237 ± 40
A PF	10	413 ± 45	230 ± 22
A	9	578 ± 47***	260 ± 25
Epo-A	9	645 ± 139***	341 ± 69**
Experiment II			
A DEF	6	473 ± 49	282 ± 48
RA PF	6	480 ± 37	276 ± 39
RA	6	603 ± 42**	284 ± 26

Values are means ± SD. A DEF = Vitamin-A-deficient; A = retinyl acetate; PF = pair-fed control; RA = retinoic acid; Epo-A = methyl 5,6-epoxyretinoate.
** $p < 0.01$, *** $p < 0.001$, compared with each A DEF group.
Retinoids (500 µg in experiment I, 100 µg in experiment II) were injected, 1 week before examination, in vitamin-A-deficient rats [122].

tion. Histological examinations of these spleens were not carried out in their study.

On the contrary, the number of splenic lymphocytes has been reported to be decreased in vitamin-A-deficient rats [54, 80, 88]. According to Nauss et al. [80], the number of splenic lymphocytes was low in vitamin-A-deficient rats in the advanced stages of vitamin A deficiency. Since there was no difference in splenic weight, on a per body weight basis, between the vitamin-A-deficient group and the pair-fed control group, they attributed the changes to the reduction in hematopoiesis in the spleens of vitamin-A-deficient animals. However, in our study, the increase in the number of splenic lymphocytes by injecting retinol or retinoic acid into vitamin-A-deficient rats was only observed when they were fed ad libitum and no difference in the number was found when food intake in the injected group was paired with that in the vitamin-A-deficient group [122].

Histological examination of the spleen from vitamin-A-deficient rats revealed a decrease in the ratio of white pulp to red pulp, small lymphocytes were confined in the attenuated periarterial lymphatic sheath, and no

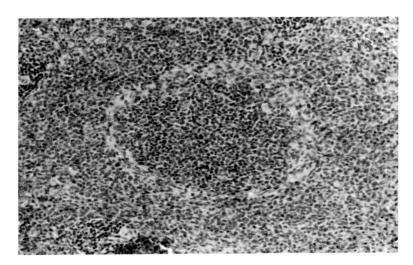

Fig. 1. Microscopic feature of spleen from a vitamin-A-deficient rat. The only rounded area found in this experiment was packed tightly together by cytoplasm-scarce lymphocytes with densely to moderately stained nucleus. This area was surrounded by a poorly stained marginal zone. Periarterial lymphatic sheath lies in the upper left corner [122].

germinal center was found in cases of the advanced vitamin A deficiency (fig. 1). Furthermore, the administration of either retinyl acetate or retinoic acid can act in the spleen to develop paracortical areas and germinal centers. In keeping with this, an increase in non-T cells was also found after the injection of retinyl acetate (table 2). Similar results were obtained in the study in which an additional dose of retinyl acetate was given in mice. Katz et al. [49] reported that an additional dose of retinyl acetate (1.0 mmol/kg diet) resulted in the expansion of the splenic marginal zone. They suggested that this finding might be attributed to an increase in the number of accessory cells, such as macrophages and dendritic cells. These findings are interesting because germinal centers were largely absent at postmortem findings in children with protein energy malnutrition. Germinal centers were also absent in the germ-free animals. The development of germinal centers is important in combating infectious diseases in states of malnutrition and vitamin A seems to play an important role in maintaining this ability. Consistent with this, it has been demonstrated in germ-free rats that life can be maintained actually without vitamin A [91].

Table 2. The reactivity of spleen cells to W3/13 antibody

Group	n	Reactivity	Cell count/spleen, $\times 10^6$
A DEF	11	+	114.9 ± 37.5
A	11	+	128.1 ± 28.2
A DEF	11	−	45.3 ± 14.4
A	11	−	$63.5 \pm 25.0*$

Values are means ± SD. See table 1 for abbreviations.
* $p < 0.05$ compared with A DEF group.
In A group, 100 µg of retinyl acetate was supplemented in vitamin-A-deficient rats 1 week before counting splenic cells. W3/13 positive cells express T-cells and negative cells express non-T cells [122].

C. Lymph Nodes and Peripheral Lymphocytes

Lymph nodes have also been shown to exhibit atrophy and peripheral lymphocytes have revealed reduced counts in vitamin-A-deficient animals [78, 80]. Takagi and Nakano [110] further observed that antigen stimulated trapping of lymphocytes in local lymph nodes was also decreased in vitamin-A-deficient rats. With respect to the effect of vitamin A administration, Taub et al. [112] have reported in their study on adjuvant material that, after a footpad injection of 0.5 mg of retinol, paracortical expansion and germinal center formation were seen at popliteal nodes in mice. Lymphopenia in the vitamin A rat can also be reversed by retinoic acid [64]. Mark et al. [64] reported that the supplementation of 20 µg alltrans-retinoic acid can reverse depression in the number of circulating lymphocytes in vitamin-A-deficient rats and impaired transformation responses to the mitogens, such as concanavalin A and *E. coli* lipopolysaccharide.

Other than the increased risk to diseases and mortality, little is known on the pathology of lymphoid organs in human vitamin A deficiency. Recently, West et al. [118] investigated the influence of vitamin A on human growth in Indonesia. Children assigned to receive vitamin A gained more body mass among boys. Furthermore, the increased risk to disease in vitamin A deficiency appeared to be more predominant among boys than girls. This may be suggestive too because Pasatiempo et al. [88] have observed splenic atrophy in male Lewis rats. However, it has yet to be determined to what degree gender differences affect to the lymphoid organs in human vitamin A deficiency.

IV. Immune Competence and Retinoids

Vitamin A deficiency profoundly affect lymphoid organs as described above. Consequently, the immune function is also impaired by vitamin A deficiency. Even in cases, in which no atrophy of lymphoid organs was observed, impaired immune functions were found by various immunological examinations. On the other hand, the administration of retinoids appears to enhance immunity not only in vitamin-A-deficient states, but also in states, in which experimental animals are fed commercial diets. From this point of view, retinoids have been termed an adjuvant material.

A. Cellular Immunity
1. Delayed Type Hypersensitivity

Delayed type hypersensitivity is impaired in vitamin A deficiency. Smith et al. [101] indicated that delayed type hypersensitivity response to dinitro-fluorobenzene was decreased in vitamin-A-deficient mice. This was significant since no difference in growth was found between the vitamin-A-deficient group and the controls. Thus, they suggested that impaired cell-mediated immunity is a primary consequence of vitamin A deficiency. On the other hand, in vitamin-A-deficient rats, we found that reactivity to phytohemagglutinin was increased by the injection of retinoids, such as retinyl acetate, retinoic acid and epoxyretinoic acid [table 3]. In human studies, Jayalakshmi and Gopalan [44] reported that children with vitamin A deficiencies had decreased responses to BCG. Although impaired delayed type hypersensitivity is also observed in other types of malnutrition, the increase in delayed type hypersensitivity was significant when compared with the pair-fed controls in our study. Thus, it seems appropriate to examine vitamin A status, when reduced delayed type hypersensitivity responses are observed. Even when no signs of vitamin A deficiency exist, the administration of retinoids may enhance reactivity by improving the body levels of vitamin A. Furthermore, vitamin A may act as an adjuvant material to enhance delayed type hypersensitivity in mice. Athanasiades [3] reported that cell-mediated immunity to intradermal SRBC was stimulated by a single simultaneous injection of 150 μg vitamin A. Miller et al. [68] observed that mice fed a diet enriched in vitamin A (238,500 IU/kg diet) exhibited enhanced contact sensitivity to oxazolone. In humans, Micksche observed that cancer patients treated with retinoids developed an increase in skin reaction to various antigens [66, 67].

Table 3. Effect of retinoid injection on phytohemagglutinin skin test

Experiment I			Experiment II		
group	n	diameter, mm	group	n	diameter, mm
A DEF	9	4.4 ± 2.0	A DEF	6	5.7 ± 1.4
A PF	9	7.4 ± 1.7**	RA PF	6	7.2 ± 1.1
A	9	7.6 ± 2.0**	RA	6	7.3 ± 1.0
Epo-A	7	8.9 ± 2.5**			

Values are means ± SD. See table 1 for abbreviations.
** $p < 0.01$ compared with each A DEF group.
1 µg/day of retinoid was injected into vitamin-A-deficient rats for 7 days. The reactivities to phytohemagglutinin are expressed as mean diameter of erythematous indurated lesions [122].

2. Skin Graft Rejection

Skin graft rejection or host-versus-graft reaction also seems to be affected by the administration of retinoids. Jurin and Tannock [46] reported that skin grafts from male C57BL/6 mice, transplanted onto isologous female mice, were rejected significantly faster when the transplanted mice were injected 250 IU/g weight/day of vitamin A for the 5 days preceding and following grafting. The same accelerated rejection was also observed after the injection of retinoic acid in mice [31]. Malkovsky et al. [63] reported that mice fed a diet containing 0.5 g/kg diet of retinylacetate responded to a suboptimal dose of semiallogeneic cells in a positive host-versus-graft reaction. They further found that, by injecting lymphoic cells into the mice fed a conventional diet, the host-versus-graft reaction in the control mice was also enhanced. Thus, they suggested that helper or inducer cells were involved in this mechanism.

3. Tumor and Cell Mediated Immunity

Since the administration of retinoids can potentiate skin graft rejection, it seems reasonable to assume that retinoids can also modify the host response to transplantable tumors. Thus, several reports have demonstrated that retinoids can inhibit the growth of transplantable tumors, such as murine melanoma, in vivo [29, 30, 108]. Recently, Forni et al. [32] have shown reductions of three transplantable tumors (spontaneous TS/A mammary adenocarcinoma, 3-methylcholanthrene-induced CE-2 fibrosarcoma and WEHI-164.1 fibrosarcoma induced by the same agent), when BALB/c mice were maintained on a diet supplemented 200 to 1,000 IU/day of vitamin A. Furthermore, there was a linear relationship between the amount of supplemental vitamin A and the extension of tumor latency or survival time. In the same study, they also found a significant increase both in proliferative responses of spleen cells to Con A and LPS and in the production of lymphokines, such as interleukin-2 and interferon-γ. From these results, the authors suggested that the resistance to tumor growth may be due to enhanced cell mediated immunity, such as killer T cell activity.

On the induction of cytotoxic cells, Dennert and Lotan [22] found in mice that the induction of cell-mediated cytotoxicity to allogeneic tumor cells was stimulated at least tenfold by a 25–300 μg/day injection of retinoic acid. They further indicated that stimulated cytotoxic activity may be effected by killer T cells and that the stimulation is dependent on antigens. The authors suggested that the retinoid stimulated cells are specific killer T cells [23, 24, 57]. Recently, Tomita et al. [114] have also shown in mice the enhancement of cytotoxic tumor rejection, mediated by T cells. However, it is still to be determined the precise mechanism of retinoid effects on killer T cells.

Natural killer cells also appear to be affected by retinoids. Irradiated mice that lose natural killer cell activity can be protected from developing leukemia either by reconstitution with cloned NK cells or by retinoid injection [24].

B. Humoral Immunity

Although accumulating evidence indicates the impaired humoral response in vitamin A deficiency, there are some difficulties in determining humoral immunity in vitamin-A-deficient animals at the basal level. Vitamin A deficiency also results in anorexia and changes in protein metabolism. Serum albumin decreases and hematocrits increase during the progressive stage of vitamin A deficiency. Thus, these findings suggest that plasma volume may change in the vitamin-A-deficient animals even when compared with pair-fed controls. As a consequence, the concentrations of humoral factors in the plasma may also be affected by these changes. When retinoids are injected into vitamin-A-deficient rats, it seems that serum albumin increases sooner, at least, than immunoglobulin. In our study, only the injection of epoxyretinoic acid, which did not affect serum albumin levels in vitamin-A-deficient rats, elevated serum IgG level in vitamin-A-deficient rats [120, 122].

However, an impaired humoral immune response to various antigens has been indicated in vitamin A deficiency. With respect to antibody titers after challenge, Krishnan et al. [54] reported that the response to SRBC immunization showed a hemagglutinin titer depressed by 50% in severely vitamin-A-deficient rats. Furthermore, responses to diphtheria and tetanus toxoids were also reduced in the vitamin-A-deficient group, when compared with pair-fed controls. Chandra and Au [14] also reported that, although no reduction in plaque-forming cells was found in vitamin-A-deficient rats, plaque-forming cell responses per spleen were reduced in vitamin-A-deficient rats, after the 5th day postimmunization. However, there was a fundamental difference in their results on spleen weights and cell counts. In Krishnan's report, severe atrophy in the thymus and spleen was seen in vitamin-A-deficient rats, but no changes were observed in the splenic weight or cellularity in the report of Chandra and Au when compared with pair-fed controls. Differences in the stages of vitamin A deficiency may partly explain the differences between these results. Recently, Pasatiempo et al. [88] indicated gender differences in the Lewis rat. They found in vitamin-A-depleted Lewis rats that spleen weight and cellularity, and plaque-forming cell responses were reduced in male rats, but no difference in spleen weight and cellularity was seen in female rats, when compared with pair-fed controls. In vitamin-A-deficient chicks, Panda and Combs [87] also observed a significant decrease in antibody titers after challenges. On the effect of supplemental retinoid administration, Jurin and Tannock [46] observed that the intraperitoneal injection of vitamin A (250 IU/g day) in mice led to a large increase in the hemoagglutinin titer, when injected simultaneously with sheep red blood cells. Thus, it was suggested that vitamin A may not act only to maintain the immune response, but may also act as an adjuvant to enhance the immune response.

With respect to immunoglobulin response, Smith et al. [101] reported that vitamin-A-deficient mice produced less IgG and IgM antibody to hemocyanin throughout the course of response. Since there was no difference in the response kinetics, they suggested that this fact may have been due to the proportionally fewer antigen specific B cells, rather than a reduced IgM secretion rate per cell. They also found some difference between IgM and IgG in vitamin-A-deficient mice. According to their report, the IgG response was more predominantly impaired than the IgM response in vitamin-A-deficient mice [100, 101]. These results are compatible with an observation of the increase in serum IgG levels after epoxyretinoid injection into vitamin-A-deficient rats. They suggested from these data that helper/inducer T cell defects in vitamin-A-deficient mice might be involved in this mechanism.

As for other immunoglobulins investigated, secretory IgA is impaired in vitamin-A-deficient rats. Sirisinha et al. [97] have demonstrated that, although there were no differences in serum IgG or IgA levels in their experiment, the secretory IgA levels in the intestinal fluid of vitamin-A-deficient rats were significantly lower than in controls. They further found that biliary secretion of IgA was also impaired in vitamin-A-deficient rats [90].

With regard to the adjuvant properties of retinol, Bryant and Barnett [13] reported that IgE response to ovalbumin was enhanced by intraperitoneal injections of retinol (1,000–9,000 IU/animal). On the other hand, the complement levels were reported to be elevated in vitamin A deficiency when examined by CH_{50} in rats [61]. However, it is still to be determined to what extent vitamin A has an effect on the complement system.

C. Phagocytosis and Phagocytes

Ongsakul et al. [85] have indicated that phagocytic activity and blood clearance of bacteria are impaired in vitamin-A-deficient rats. They further found that the phagocytic activity of polymorphonuclear leukocytes was similarly affected. On the effect of supplemental doses of vitamin A, Moriguchi et al. [71] reported that very high intakes of vitamin A can increase the phagocytic activity of peritoneal macrophages and potentiate the production of interleukin-1 in mice. Then, Trechsel et al. [115] supported this hypothesis concerning the stimulation of interleukin-1 at the physiological level by retinoic acid. They also found stimulation of interleukin-3. Recently, Dillehay et al. [26] have reported in their in vitro study, using a mouse macrophage cell line, that phagocytosis of IgG-sensitized bovine erythrocytes increased significantly with very low levels (10^{-10} M) of retinoic acid. They also found potentiation of interleukin-1 activity at a retinoic acid concentration of 10^{-8} M. Using human macrophages obtained from activated peripheral blood monocytes of preschool children, Bhaskaram et al. [6] found a significant potentiation of interleukin-1 production after the oral administration of 100,000 IU vitamin A, regardless of the initial serum vitamin A level. However, no change in cytotoxic function was observed after the administration of vitamin A. Considering the route and dose of vitamin A in these experiments, oral vitamin A may stimulate interleukin production via retinoic acid. Retinoid seems to act, at least, in two different ways to potentiate phagocytic functions.

Regarding the tumoricidal activity of macrophages, Tachibana et al. [109] have studied the effect of orally administered vitamin A at a dosage of 100–500 IU per gram body weight for 4 consecutive days in the rat. They found an increase in cytotoxicity, mediated by alveolar macrophages, against a syngeneic mammary adenocarcinoma cell line. The same effect was obtained in their in vitro study and retinoic acid also induced tumoricidal activity. Moriguchi et al. [72] carried out in vitro study with human monocytes. Cytotoxic monocytic activity was induced following incubation with $10^{-8} M$ retinol for 24 h.

D. Regulation of Lymphocyte Transformation

Lymphocyte transformation has also been examined in vitamin A deficiency and the level of mitogen response was impaired in vitamin A deficiency. The response to concanavalin A or phytohemagglutinin was depressed in vitamin-A-deficient rats [14, 78]. The response to lipopolysaccharide was also markedly reduced in vitamin-A-deficient animals [88]. Thus, the blast formation to both T-cell mitogen and B-cell mitogen may be depressed in vitamin A deficiency [108]. On the other hand, supplementary retinoids may increase mitogen responsiveness [29, 55, 94, 95]. Recently, it has been indicated that retinoic acid upregulates interleukin-2 receptors on activated human thymocytes [96]. It is possible that retinoic acids may act on activated lymphocytes to modulate interleukin-2-dependent immune responses.

V. Mechanism of Action

Although intensive investigations have recently been performed, the precise mechanism of action of vitamin A has not yet been determined. However, recent progress in biological sciences indicates a profound effect of retinoid on cell. This section is designed to summarize current progress on two points. The first point involves looking for active metabolites of vitamin A and the other is to investigate the mechanism of its action on immune system.

A. Active Metabolites of Vitamin A

Although retinal is the active metabolite in vision, it is hard to suppose that this metabolite is the active metabolite in all other physiological functions. In 1960, Dowling and Wald [27] reported that retinoic acid was as active as vitamin A in growth of rats. However, these rats maintained on retinoic acid, instead of vitamin A, were severely night-blind. Malathi et al. [62] found that retinoic acid was 141% as biologically active as vitamin A acetate, when injected intraperitoneally. In 1967, 5,6-epoxyretinoic acid was implicated as more biologically active retinoid [45]. These retinoids were identified as physiological metabolites [59, 76, 77]. On the other hand, Zile et al. [123] reported that 5,6-epoxyretinoic acid exhibited poor activity. This was also confirmed in our study [120, 122]. Although this compound restored some of impaired immunity in vitamin-A-deficient rats, the growth promoting effect of 5,6-epoxyretinoic acid was not seen in vitamin-A-deficient rats [122].

Retinoic acid, on the other hand, appears to be more biologically active than vitamin A. When the action of retinylidene acetic acid was examined [121], it was noticed that the growth of vitamin-A-deficient rats was faster and better in animals supplemented with retinoic acid than in animals supplemented with vitamin A acetate (fig. 2).

The effect of retinoids on immune functions also seems to be different between these metabolites. Although retinol, retinal and epoxyretinoic acid exhibit different effects from retinoic acid on immune function, retinoic acid is generally more active than these metabolites [24, 122]. Furthermore, retinoic acid has cellular binding proteins in many tissues, including the reproductive system. Recently, it was discovered that retinoic acid can act as a morphogen in developing chick bud [99, 113]. A nucleic receptor for retinoic acid has also been identified [89]. Thus, it is assumed that retinoic acid may act in the host-defense system as a chemopreventive agent for infections and cancers. From this point of view, it is interesting that in macrophages, within minutes of exposure to retinoic acid, the accumulation of tissue transglutaminase mRNA is induced by cDNA clones for this enzyme [15]. The same effect was also observed in human promyelocytic leukemia cells [21, 75]. Thus, the authors suggested that retinoic acid may regulate gene expression by both transcriptional and posttranscriptional mechanisms. Tissue transglutaminase, a calcium-dependent enzyme, catalyzes the cross-linking of specific membrane associated proteins, such as fibronectin [74]. Considering the rapidness of this accumulation, it may be a fundamental step in the

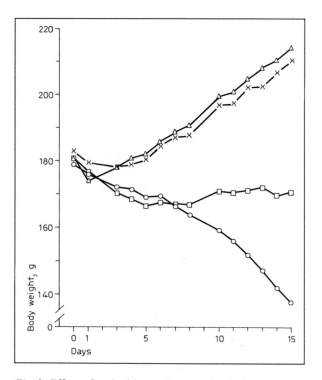

Fig. 2. Effect of retinoids on the growth of vitamin-A-deficient rats. Each mark indicates mean body weight and horizontal axis indicates time after starting retinoid supplementation. ○ = Vitamin-A-deficient (n = 8); □ = retinylidene acetic acid (n = 10); △ = retinoic acid (n = 10); × = retinyl acetate (n = 9). Retinoids (4 mg/kg diet) were supplemented in vitamin-A-deficient rats. The growth of rats given retinoic acid is seemed to respond sooner and better than the rats given retinyl acetate [121].

effect of retinoic acid on immune competence. However, it is also possible that other metabolites are also involved in the mechanism of this effect.

B. Retinoid Mechanism in the Host-Defense System

As reviewed above, vitamin A and retinoids are seemed to have multiple effects in host-defense systems. Retinoids not only regulate enzyme synthesis and gene expression, but also membrane function by influencing glycoproteins. Thus, it is difficult to explain the mechanisms of retinoid action. As indicated above, retinoic acid rapidly induces tissue transglutaminase activity in macrophages [16, 65]. Recently, Mehta et al. [65] have reported that the cytostatic activity in mouse peritoneal macrophages is negatively correlated with transglutaminase activity. They suggested the association of retinoid-induced expression of transglutaminase with the inhibition of cytostatic activation. Tissue transglutaminase may also regulate ornithine decarboxylase activity, which is induced with tumor promotion and inhibited by retinoid. At least, macrophages appear to be an important site of action for retinoid, because at the physiological levels of $10^{-8}-10^{-10}$ M retinoic acid can successfully increase the phagocytic activity and the potentiation of interleukin-1 [26]. Interleukin-1 is necessary for T-cell expansion. 10^{-8} M of retinoic acid can also enhance human thymocyte response, which is dependent on interleukin-2 [96]. The authors suggested in the report that retinoic acid may enhance the response by increasing interleukin-2 receptor expression. Recently, the same level of retinoic acid was reported to increase steady levels of μ- and κ-chain mRNA in B-cell hybridomas, which have a characteristic of resting B cell [93]. Thus, retinoid may modulate gene expression other than that of tissue transglutaminase in macrophages.

Recently, it was proposed that retinoic acid can affect the protein kinase C cascade system [19, 43]. This system is supposed to provide a signal-transducing system for cell responses [50, 84, 117]. In short, this system involves the following events. When a transmembrane signal activates the membrane phospholipase C, second messengers, such as diacylglycerol and inositol triphosphate, are generated [5]. Then, diacylglycerol induces the translocation of protein kinase C [50]. Inositol triphosphate, on the other hand, mobilizes calcium to increase free calcium concentration in the cell. This enzyme is activated by Ca^{2+} and phosphatidylserine [52]. The membrane activated protein kinase C may transmit a signal to increase gene expression or to modulate membrane function. The precise role of this cascade system is presently under investigation. However, this system appears to be involved in both carcinogenic process and immune responses. This system is activated by phorbol ester which acts as a potent tumor promoter. The activation of this system may increase the expression of interleukin-2 receptor and its production to cause a proliferative response in T cells [43]. Since retinoids influence both the carcinogenic process and immune response, it is possible for retinoids to act in this system.

From this point of view, it is interesting that retinoic acid acts on T cells by increasing interleukin-2 receptors [96]. Recently, Cope [19] has shown that cellular binding proteins of retinol and retinoic acid are substrates for protein kinase C and that their holo-forms inhibit the activity of protein kinase C. Isakov [43] demonstrated that T cell derived protein kinase C activity was potentiated by retinoic acid, but was inhibited by retinal. Bosma and Sidell [12] reported the inhibition of Ca^{2+} currents and cell proliferation in a B-lymphocyte cell line. Although the concentration of retinoids, used in some of their experiments, seemed higher than physiological levels, it is possible that the effect of retinoids on the host defense system may be mediated by the protein kinase C cascade system. Further studies will be needed to determine the degree of regulation involved in the physiological role of vitamin A.

References

1   Anonymous: Diet, nutrition, and cancer. Committee on Diet, Nutrition, and Cancer, Assembly of Life Sciences, National Research Council (National Academy Press, Washington 1982).
2   Anzano, M.A.; Lamb, A.J.; Olson, J.A.: Growth, appetite, sequence of pathological signs and survival following the induction of rapid synchronous vitamin A deficiency in the rat. J. Nutr. *109:* 1419–1431 (1979).
3   Athanasiades, T.J.: Adjuvant effect of vitamin A palmitate and analog on cell mediated immunity. J. natn. Cancer Inst. *67:* 1153–1156 (1981).
4   Barclay, A.J.G.; Foster, A.; Sommer, A.: Vitamin A supplements and mortality related to measles: a randomized clinical trial. Br. med. J. *294:* 294–296 (1987).
5   Berridge, M.J.: Inositol triphosphate and diacylglycerol as second messengers. Biochem. J. *220:* 45–60 (1984).
6   Bhaskaram, P.; Sharada, K.; Sivakumar, B.; Rao, K.V.; Nair, M.: Effect of iron and vitamin A deficiencies on macrophage function in children. Nutr. Res. *9:* 35–45 (1989).
7   Bieri, J.G.; McDaniel, E.G.; Rogers, W.E.: Survival of germfree rats without vitamin A. Science *163:* 574–575 (1968).
8   Biesel, W.R.: Role of Nutrition in immune system disease. Compr. Ther. *13:* 13–19 (1987).
9   Biesel, W.R.: Single nutrients and immunity. Am. J. clin. Nutr. *35:* suppl, pp. 417–468 (1982).
10 Bjelke, E.: Dietary vitamin A and human lung cancer. Int. J. Cancer *15:* 561–565 (1975).
11 Bondi, A.; Sklan, D.: Vitamin A and carotene in animal nutrition. Prog. Food Nutr. Sci. *8:* 165–191 (1984).
12 Bosma, M.; Sidell, N.: Retinoic acid inhibits Ca^{2+} currents and cell proliferation in a B-lymphocyte cell line. J. cell. Physiol. *135:* 317–323 (1988).
13 Bryant, R.L.; Barnett, J.B.: Adjuvant properties of retinol on IgE production in mice. Int. Archs Allergy appl. Immunol. *59:* 69–74 (1979).

14 Chandra, R.K.; Au, B.: Single nutrient deficiency and cell mediated immune responses. III. Vitamin A. Nutr. Res. *1:* 181–185 (1981).

15 Chiocca, E.A.; Davies, P.J.A.; Stein, J.P.: The molecular basis of retinoic acid action: transcriptional regulation of tissue transglutaminase gene expression in macrophages. J. biol. Chem. *263:* 11584–11589 (1988).

16 Chiocca, E.A.; Davies, P.J.A.; Stein, J.P.: Regulation of tissue transglutaminase gene expression as a molecular model for retinoid effects on proliferation and differentiation. J. Cell Biochem. *39:* 293–304 (1989).

17 Cohen, B.E.; Elin, R.J.: Vitamin A induced non-specific resistance to infection. J. infect. Dis. *129:* 597–600 (1974).

18 Cohen, S.M.; Whittenberg, J.F.; Bryan, G.T.: Effects of avitaminosis A and hypervitaminosis A on urinary bladder carcinogenicity of N-[4-(5-nitro-2-furyl)-2-thiazolyl]formamide. Cancer Res. *17:* 2334–2339 (1976).

19 Cope, F.O.; Howard, B.D.; Boutwell, R.K.: The in vitro characterization of the inhibition of mouse brain kinase-C by retinoids and their receptors. Experientia *42:* 1023–1027 (1986).

20 Davies, R.E.: Effect of vitamin A on 7,12-dimethlbenz(α)-anthracene-induced papillomas in rhino mouse skin. Cancer Res. *27:* 237–241 (1967).

21 Davies, P.J.A.; Murtaugh, M.P.; Moore, W.T.; Johnson, G.S.; Lucas, D.: Retinoic-acid induced expression of tissue transglutaminase in human promyelocytic leukemia (HL-60) cells. J. biol. Chem. *260:* 5166–5176 (1985).

22 Dennert, G.; Lotan, R.: Effect of retinoic acid on the immune system: stimulation of T killer cell induction. Eur. J. Immunol. *8:* 23–29 (1978).

23 Dennert, G.; Crowley, C.; Kouba, J.; Lotan, R.: Retinoic acid stimulation of the mouse killer T-cells in allogeneic and syngeneic systems. J. natn. Cancer Inst. *62:* 89–94 (1979).

24 Dennert, G.: Immunostimulation by retinoic acid; in Nugent, Clark, Retinoids, differentiation and disease. Ciba Foundation Symposium 113, pp. 117–131 (Pitman, London 1985).

25 De Sole, G.; Belay, Y.; Zegeye, B.: Vitamin A deficiency in southern Ethiopia. Am. J. Clin. Nutr. *45:* 780–784 (1987).

26 Dillehay, D.L.; Walia, A.S.; Lamon, E.W.: Effect of retinoids on macrophage function and IL-1 activity. J. Leukocyte Biol. *44:* 353–360 (1988).

27 Dowling, J.E.; Wald, G.: The biological function of vitamin A acid. Proc. natn. Acad. Sci. USA *46:* 587–608 (1960).

28 Eccles, S.A.; Barnett, S.C.; Alexander, P.: Inhibition of growth and spontaneous metastasis of syngeneic transplantable tumours by an aromatic retinoic acid analog. 1. Relationship between tumour immunogenicity and responsiveness. Cancer Immunol. Immunother. *19:* 109–114 (1985).

29 Eccles, S.A.: Effects of retinoids on growth and dissemination of malignant tumours: immunological considerations. Biochem. Pharmac. *34:* 1599–1610 (1985).

30 Eremin, O.; Ashby, J.; Rhodes, J.: Inhibition of antibody-dependent cellular cytotoxicity by retinoic acid. Int. Archs Allergy appl. Immunol. *75:* 2–7 (1984).

31 Floersheim, G.L.; Bollag, G.: Accelerated rejection of skin homografts by vitamin A acid. Transplantation *14:* 564–567 (1972).

32 Forni, G.; Sola, S.C.; Giovarelli, M.; Santoni, A.; Martinetto, P.; Vietti, D.: Effect of prolonged administration of low doses of dietary retinoids on cell-mediated immu-

nity and the growth of transplantable tumors in mice. J. natn. Cancer Inst. *76:* 527–533 (1986).

33 Fujimaki, Y.: Formation of gastric carcinoma in albino rats fed on deficient diets. J. Cancer Res. *10:* 469–477 (1926).

34 Gershwin, M.E.; Beach, R.S.; Hurley, L.S.: Nutrition and Immunity; pp. 228–239 (Academic Press, New York 1985).

35 Giguere, V.; Ong, E.S.; Segui, P.; Evans, R.M.: Identification of a receptor for the morphogen retinoic acid. Nature, Lond. *330:* 624–629 (1987).

36 Goodman, D.S.: Vitamin A and retinoids in health and disease. New Engl. J. Med. *310:* 1023–1031 (1984).

37 Green, H.N.; Mellanby, E.: Vitamin A as an anti-infective agent. Br. med. J. *ii:* 691–692 (1928).

38 Green, H.N.; Mellanby, E.: Carotene and vitamin A: The anti-infective action of carotene. Br. J. exp. Path. *11:* 81–89 (1930).

39 Hayes, K.C.: On the pathophysiology of vitamin A deficiency. Nutr. Rev. *29:* 3–6 (1971).

40 Henneckens, C.H.; Mayrent, S.L.; Willett, W.: Vitamin A, carotenoids, and retinoids. Cancer *58:* Suppl., pp. 1837–1841 (1986).

41 Hirayama, T.: Diet and cancer. Nutr. Cancer *1:* 67–81 (1979).

42 Hirayama, T.: A large scale cohort study on cancer risk reducing effects of green-yellow vegetable consumption. Int. Symp. Princess Takamatsu Cancer Res. Fund, vol. 16, pp. 41–53 (1985).

43 Isakov, N.: Regulation of T-cell-derived protein kinase C activity by vitamin A derivatives. Cell. Immunol. *115:* 288–298 (1988).

44 Jayalakshmi, V.T.; Gopalan, C.: Nutrition and tuberculosis. Part I. An epidemiological study. Indian J. med. Res. *46:* 87–92 (1958).

45 John, K.V.; Lakshaman, M.R.; Cama, H.R.: Preparation, properties, and metabolism of 5,6-monoepoxyretinoic acid. Biochem. J. *103:* 539–543 (1967).

46 Jurin, M.; Tannock, I.F.: Influence of vitamin A on the immunologic response. Immunology *23:* 283–287 (1972).

47 Kark, J.D.; Smith, A.H.; Schwitzer, B.R.; Hames, C.G.: Serum vitamin A (retinol) and cancer incidence in Evans County, Georgia. J. natn. Cancer Inst. *66:* 7–16 (1981).

48 Kark, J.D.; Smith, A.H.; Hames, C.G.: Serum retinol and the inverse relationship between serum cholesterol and cancer. Br. med. J. *284:* 152–154 (1982).

49 Katz, D.R.; Drzymala, J.A.; Turton, R.M.; Hicks, R.M.; Hunt, H.R.; Palmer, L.; Malkovsky, M.: Regulation of accessory cell function by retinoids in murine immune responses. Br. J. exp. Path. *68:* 343–350 (1987).

50 Kikkawa, U.; Nishizuka, Y.: The role of protein kinase C in transmembrane signalling. Ann. Rev. cell. Biol. *2:* 149–178 (1986).

51 Kim, H.Y.; Wolf, G.: Vitamin A deficiency alters genomic expression for fibronectin in liver and hepatocytes. J. biol. Chem. *262:* 365–371 (1987).

52 Kuno, M.; Garner, P.: Ion channels activated by inositol 1,4,5-triphosphate in plasma membrane of human T-lymphocytes. Nature, Lond. *326:* 301–304 (1987).

53 Krinsky, N.I.: Carotenoids and cancer in animal models. J. Nutr. *119:* 123–126 (1989).

54 Krishnan, S.; Bhuyan, U.N.; Talwar, G.P.; Ramalingaswami, V.: Effect of vitamin A

and protein-calorie undernutrition on immune response. Immunology 27: 383–392 (1974).

55 Lippman, S.M.; Kessler, J.F.; Meyskens, F.L.: Retinoids as preventive and therapeutic anticancer agents (part I). Cancer Treatm. Rep. 71: 391–405 (1987).

56 Lippman, S.M.; Kessler, J.F.; Meyskens, F.L.: Retinoids as preventive and therapeutic anticancer agents (part II). Cancer Treatm. Rep. 71: 493–515 (1987).

57 Lotan, R.; Dennert, G.: Stimulatory effect of vitamin A analogs on induction of cell mediated cytotoxicity in vivo. Cancer Res. 39: 55–58 (1979).

58 McCollum, E.V.: The supplementary dietary relationships among our natural foodstuffs. J. Am. med. Ass. 68: 1379–1386 (1917).

59 McCormick, A.M.; Napoli, J.L.; Schnoes, H.K.; Deluca, H.F.: Isolation and identification of 5,6-epoxyretinoic acids: A biologically active metabolite of retinoic acid. Biochemistry 17: 4085–4090 (1978).

60 MacMurray, D.N.: Cell-mediated immunity in nutritional deficiency. Prog. Food Nutr. Sci. 8: 193–228 (1984).

61 Madjid, B.; Sirisinha, S.; Lamb, A.J.: The effect of vitamin A and protein deficiency on complement levels in rats. Proc. Soc. exp. Biol. Med. 158: 92–95 (1978).

62 Malathi, K.P.; Rao, S.; Sastry, P.S.; Ganguly, J.: Studies on metabolism of vitamin A. 1. The biological activity of vitamin A acid in rats. Biochem. J. 87: 305–311 (1963).

63 Malkovsky, M.; Edwards, A.J.; Hunt, R.; Palmer, L.; Medawar, P.B.: T-cell-mediated enhancement of host-versus-graft reactivity in mice fed a diet enriched in vitamin A acetate. Nature, Lond. 302: 338–340 (1983).

64 Mark, D.A.; Baliga, B.S.; Suskind, R.M.: All-trans retinoic acid reverses immune-related hematological changes in the vitamin A deficient rat. Nutr. Rep. int. 28: 1245–1252 (1983).

65 Mehta, K.; Claringbold, P.; Lopez-Berestein, G.: Suppression of macrophage cytostatic activation by serum retinoids: a possible role for transglutaminase. J. Immunol. 138: 3902–3906 (1987).

66 Micksche, M.; Cerni, C.; Kokron, O.; Tischer, R.; Wrba, H.: Stimulation of immune response in lung cancer patients by vitamin A therapy. Oncology 34: 234–238 (1977).

67 Micksche, M.: Immunologie und Immunotherapie des Lungenkarzinoms: Experimentelle und klinische Untersuchungen. Wien. klin. Wschr. 90: suppl., pp. 1–28 (1978).

68 Miller, K.; Maisey, J.; Malkovsky, M.: Enhancement of contact sensitization in mice fed a diet enriched in vitamin A acetate. Int. Archs Allergy appl. Immunol. 75: 120–125 (1984).

69 Milton, R.C.; Reddy, V.; Naidu, A.N.: Mild vitamin A deficiency and childhood morbidity – an Indian experience. Am. J. clin. Nutr. 46: 827–829 (1987).

70 Moon, R.C.; McCormick, D.L.; Mehta, R.G.: Inhibition of carcinogenesis by retinoids. Cancer Res. 43: 2469–2475 (1983).

71 Moriguchi, S.; Werner, L.; Watson, R.R.: High dietary vitamin A (retinyl palmitate) and cellular immune function in mice. Immunology 56: 169–177 (1985).

72 Moriguchi, S.; Kohge, M.; Kishino, Y.; Watson, R.R.: In vitro effect of retinol and 13-cis retinoic acid on cytotoxity of human monocytes. Nutr. Res. 8: 255–264 (1988).

73 Muhilal, Permeisih, D.; Idjradinata, Y.R.; Muherdiyantiningsih; Karyadi, D.: Vitamin A-fortified monosodium glutamate and health, growth, and survival of children: a controlled field trial. Am. J. clin. Nutr. *48:* 1271–1276 (1988).

74 Murtaugh, M.P.; Davies, P.J.A.: Transglutaminase expression and fibronectin cross-linking on the cell surface of peritoneal macrophages. J. Cell Biol. *99:* 326a (1984).

75 Murtaugh, M.P.; Dennison, O.; Stein, J.P.; Davies, P.J.A.: Retinoic acid-induced gene expression in normal and leukemic myeloid cells. J. exp. Med. *163:* 1325–1330 (1986).

76 Napoli, J.L.; Race, K.R.: The biosynthesis of retinoic acid from retinol by rat tissue in vitro. Archs Biochem. Biophys. *255:* 95–101 (1987).

77 Napoli, J.L.; Harris, S.; Shields, C.; Cummings, C.: Retinol metabolism in LLC-PK1 Cells. J. biol. Chem. *261:* 13592–13597 (1987).

78 Nauss, K.M.; Mark, D.A.; Suskind, R.M.: The effect of vitamin A deficiency on the in vitro cellular immune response of rats. J. Nutr. *108:* 1815–1823 (1979).

79 Nauss, K.M.; Anderson, C.A.; Conner, M.W.; Newberne, P.M.: Ocular infection with herpes simplex virus (HSV-1) in vitamin A-deficient and control rats. J. Nutr. *115:* 1300–1315 (1985).

80 Nauss, K.M.; Phua, C.-C.; Ambrogi, L.; Newberne, P.M.: Immunological changes during progressive stages of vitamin A deficiency in the rat. J. Nutr. *115:* 909–918 (1985).

81 Nettesheim, P.; Snyder, C.; Williams, M.L.; Cone, M.V.; Kim, J.C.: Effect of vitamin A on lung tumor induction in rats. Proc. Am. Ass. Cancer Res. *16:* 54 (1975).

82 Nettesheim, P.; Cone, M.V.; Snyder, C.: The influence of retinyl acetate on the postinitiation phase of preneoplastic lung nodules in rats. Cancer Res. *36:* 996–1002 (1976).

83 Newberne, P.M.; Rogers, A.E.: Rat colon carcinomas associated with aflatoxin and marginal vitamin A. J. natn. Cancer Inst. *50:* 439–448 (1973).

84 Nishizuka, Y.: Studies and perspective of protein kinase C. Science *233:* 305–312 (1986).

85 Ongsakul, M.; Sirishinha, S.; Lamb, A.J.: Impaired blood clearance of bacteria and phagocytic activity in vitamin A deficient rats. Proc. Soc. exp. Biol. Med. *178:* 204–208 (1985).

86 Paganini-Hill, A.; Chao, A.; Ross, R.K.; Henderson, B.E.: Vitamin A, beta-carotene, and the risk of cancer: A prospective study. J. natn. Cancer Inst. *79:* 443–448 (1987).

87 Panda, B.; Combs, G.F.: Impaired antibody production in chicks fed diet low in vitamin A. Proc. Soc. exp. Biol. Med. *113:* 530 (1963).

88 Pasatiempo, A.M.G.; Bowman, T.A.; Taylor, C.E.; Ross, A.C.: Vitamin A depletion and repletion: effects on antibody response to the capsular polysaccharide of Streptococcus pneumoniae, type III (SSS-III). Amer. J. clin. Nutr. *49:* 501–510 (1989).

89 Petkovich, M.; Brand, N.J.; Krust, A.; Chambon, P.: A human retinoic acid receptor which belongs to the family of nuclear receptors. Nature, Lond. *330:* 444–450 (1987).

90 Puengtomwatanakul, S.; Sirisinha, S.: Impaired biliary immunoglobulin A in vitamin A-deficient rats. Proc. Soc. exp. Biol. Med. *182:* 437–442 (1986).

91 Rogers, W.E.; Bieri, J.G.; McDaniel, E.G.: Vitamin A deficiency in the germfree state. Fed. Proc. *30:* 1773–1778 (1971).

92 Scrimshaw, N.S.; Taylor, C.E.; Gordon, J.E.: Interactions of nutrition and infection. WHO Monogr. Ser. No. 57, pp. 40–41, 87–96, 143–180 (1968).

93 Sherr, B.; Adelman, D.C.; Saxon, A.; Gilly, M.; Wall, R.; Sidell, N.: Retinoic acid induces the differentiation of B cell hybridomas from patients with common variable immunodeficiency. J. exp. Med. *168:* 55–71 (1988).

94 Sidell, N.; Famatiga, E.; Golub, S.H.: Augmentation of human thymocyte proliferative responses by retinoic acid. Exp. Cell Biol. *49:* 239–245 (1981).

95 Sidell, N.; Famatiga, E.; Golub, S.H.: Immunological aspects of retinoids in humans. II. Retinoic acid enhances induction of plaque forming cells. Cell. Immunol. *88:* 374–381 (1984).

96 Sidell, N.; Ramsdell, F.: Retinoic acid upregulates interleukin-2 receptors on activated human lymphocytes. Cell. Immunol. *115:* 299–309 (1988).

97 Sirisinha, S.; Darip, M.D.; Moongkarndi, P.; Ongsakul, M.; Lamb, A.J.: Impaired local immune response in vitamin A-deficient rats. Clin. exp. Immunol. *40:* 127–135 (1980).

98 Sklan, D.: Vitamin A in Human Nutrition. Prog. Food Nutr. Sci. *11:* 39–55 (1987).

99 Slak, J.M.W.: We have a morphogen! Nature, Lond. *327:* 553–554 (1987).

100 Smith, S.M.; Hayes, C.E.: Contrasting impairments in IgM and IgG responses of vitamin A-deficient mice. Proc. natn. Acad. Sci. USA *84:* 5878–5882 (1987).

101 Smith, S.M.; Levy, N.; Hayes, C.E.: Impaired immunity in vitamin A-deficient mice. J. Nutr. *117:* 857–865 (1987).

102 Sommer, A.; Tarwotjo, I.; Hussaini, G.; Susanto, D.: Increased mortality in children with mild vitamin A deficiency. Lancet *ii:* 585–588 (1983).

103 Sommer, A.; Katz, J.; Tarwotjo, I.: Increased risk of respiratory disease and diarrhea in children with preexisting mild vitamin A deficiency. Am. J. clin. Nutr. *40:* 1090–1095 (1984).

104 Sommer, A.; Tarwotjo, I.; Djunaedi, E.; West, K.P.; Loeden, A.A.; Tilden, R.; Mele, L.: Impact of vitamin A supplementation on childhood mortality. Lancet *i:* 1169–1173 (1986).

105 Sommer, A.; Tarwotjo, I.; Katz, J.: Increased risk of xerophtalmia following diarrhea and respiratory disease. Am. J. clin. Nutr. *45:* 977–980 (1987).

106 Sommer, A.: New imperatives for an old vitamin (A). J. Nutr. *119:* 96–100 (1989).

107 Sporn, M.B.; Dunlap, N.M.; Newton, D.L., et al.: Prevention of chemical carcinogenesis by vitamin A and its synthetic analogs (retinoids). Fed. Proc. *35:* 1332–1338 (1976).

108 Sporn, M.B.; Robert, A.B.; Goodman, D.S. (eds.): The retinoids, vol. 2 (Academic Press, New York 1984).

109 Tachibana, K.; Sone, S.; Tsubura, E.; Kishino, Y.: Stimulatory effect of vitamin A on tumoricidal activity of rat alveolar macrophages. Br. J. Cancer *49:* 343–348 (1984).

110 Takagi, H.; Nakano, K.: The effect of vitamin A depletion on antigen-stimulated trapping of peripheral lymphocytes in local lymphnodes of rats. Immunology *48:* 123–128 (1983).

111 Tarwotjo, I.; Sommer, A.; West, K.P.; Djunaedi, E.; Mele, L.; Hawkins, B.: Influence of participation on mortality in a randomized trial of vitamin A prophylaxis. Am. J. clin. Nutr. *45:* 1466–1471 (1987).

112 Taub, R.N.; Krantz, A.R.; Dresser, D.W.: The effect of localized injection of adjuvant material on the draining lymph node. Immunology *18:* 171–186 (1970).

113 Thaller, C.; Eichele, G.: Identification and spatial distribution of retinoids in the developing chick limb bud. Nature, Lond *327:* 625–628 (1987).

114 Tomita, Y.; Himeko, K.; Nomoto, K., et al.: Vitamin A and tumor immunity. Experientia *41:* 92–93 (1985).

115 Trechsel, U.; Evéquoz, V.; Fleisch, H.: Stimulation of interleukin 1 and 3 production by retinoic acid in vitro. Biochem. J. *230:* 339–344 (1985).

116 Wald, N.; Idle, M.; Boreham, J.; Baily, A.: Low serum-vitamin A and subsequent risk of cancer – preliminary results of a prospective study. Lancet *ii:* 813–815 (1980).

117 Weiss, A.; Imboden, J.; Hardy, K.; Manger, B.; Terhorst, C.; Stobo, J.: The role of the T3/antigen receptor complex in T-Cell activation. Ann. Rev. Immunol. *4:* 593–619 (1986).

118 West, K.P.; Djunaedi, E.; Pandji, A.; Kusdiono; Tarwotjo, I.; Sommers, A.: Vitamin A supplementation and growth: a randomized community trial. Am. J. clin. Nutr. *48:* 1257–1264 (1988).

119 Wolf, G.: Multiple functions of vitamin A. Physiol. Rev. *64:* 873–937 (1984).

120 Yamamoto, M.; Hiyama, Y.; Kitano, T.; Matsui, M.; Okada, K.: On the biological activity and immune effect of 5,6-epoxyretinoic acid. Vitamins, Kyoto *55:* 279–280 (1981).

121 Yamamoto, M.; Kitano, T.; Oma, M.; Matsui, M.: On the biological activity of retinylidene acetic acid. Presented at 40th Annual Meeting for Food and Nutrition, 1986.

122 Yamamoto, M.; Kitano, T.; Oma, M.; Hiyama, Y.; Yumoto, S.; Matsui, M.: Supplemental effect of retinoids on immune function in vitamin A deficient rats. Nutr. Res. *8:* 529–538 (1988).

123 Zile, M.H.; Inhorn, R.C.; Deluca, H.F.: The biological activity of 5,6-epoxyretinoic acid. J. Nutr. *110:* 2225–2230 (1980).

124 Zile, M.H.; Cullum, M.E.: The function of vitamin A: Current concepts. Proc. exp. Biol. Med. *172:* 139–152 (1983).

Manabu Yamamoto, MD, Department of Food and Nutrition,
Tachikawa College of Tokyo, 3-6-33 Azuma-cho, Akishima-shi, Tokyo 196 (Japan)

Simopoulos AP (ed): Selected Vitamins, Minerals, and Functional Consequences of
Maternal Malnutrition. World Rev Nutr Diet. Basel, Karger, 1991, vol 64, pp 85–108

Vitamin K Deficiency in Infancy[1]

Ichiro Matsuda, Fumio Endo, Kunihiko Motohara[2]

Department of Pediatrics, Kumamoto University Medical School, Kumamoto,
Japan

Contents

Introduction

The clinical entity of so-called 'hemorrhagic disease of the newborn'
was first established by Townsend in 1894 [1], who reported 50 infants
with bleeding during the first two weeks of life. The disease usually appears
on the second or third day of life with gastrointestinal bleeding. The mechanism underlying the disease was obscure until vitamin K was discovered
and a low level of prothrombin was detected in newborn infants. In 1929,
Dam [2] observed that when chicks were placed on a fat-free diet they bled
to death and later he reported that this was due to a deficiency of a new
unidentified chemical, which he called vitamin K. Dam et al. [3] subsequently confirmed the prothrombin deficiency in the plasma of hemor-

[1] Supported by Grant from the Ministry of Health and Welfare Japan.
[2] We thank Professor Lewis A. Barness for review of the manuscript.

rhagic chicks. Brinkhouse et al. [4] reported that prothrombin levels in normal newborn infants were lower than in adults and that the levels were reduced even further in hemorrhagic disease of the newborn. The involvement of vitamin K in the disease was also demonstrated by the facts that low prothrombin level could be increased by the administration of vitamin K [5] and prophylactic administration of the vitamin prevented the bleeding tendency in the neonatal period [6, 7]. Based on these observations, the hemorrhagic disease caused by reduced vitamin K-dependent factors was differentiated from bleeding secondary to other causes [8]. Accordingly, the Committee on Nutrition of the American Academy of Pediatrics recommended the prophylactic administration of vitamin K at birth in 1961 [9]. However, neonatal vitamin K prophylaxis remains controversial and is not uniformly practiced throughout the world. Some workers supported the idea of vitamin K prophylaxis [10–12] while others believed that it is unnecessary for healthy newborn infants [13–17].

The late neonate hemorrhagic disease, the main manifestation of which is intracranial bleeding, is another result of vitamin K deficiency, which was uncommon and until recently not reported. The disease was often found in breast-fed infants with no prophylactic vitamin K administration at birth [18]. In 1966, Goldman and Deposito [19] reported five patients with a bleeding disorder with hypoprothrombinemia that occurred after the neonatal period (range, 3–7 weeks of age) and since then many reports from various parts of the world have drawn attention to this delayed coagulopathy, as discussed later.

Recently, considerable progress has been made as to the elucidation of the biological functions of vitamin K [12, 20], due in part to the development of a skillful means of vitamin K measurement [21] and a specific method for the detection of vitamin K deficiency [22, 23]. Nevertheless, vitamin K deficiency remains one of the main worldwide causes of infant death and handicap [24, 25]. We review here our current knowledge of vitamin K to emphasize its role in the neonatal period and the necessity of vitamin K prophylaxis for all neonates.

Function of Vitamin K

Vitamin K is required for the modification and activation of a number of important proteins, so-called vitamin K-dependent proteins, which include plasma coagulation factors II, VII, IX and X, coagulation inhibitor

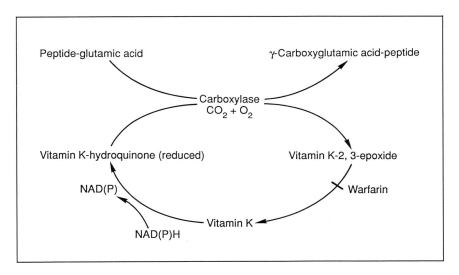

Fig. 1. Vitamin K cycle and carboxylation reaction.

In the first step, vitamin K is reduced to hydroquinone (KH_2), which is an active form, in the presence of NAPH or NADPH. KH_2 is then oxidized to vitamin K-2,3-epoxide. This reaction is associated with the carboxylation reaction, which yields GLA peptides: II, VII, IX and X, osteocalcin and other GLA peptides. Subsequently, the vitamin K epoxide is converted to native vitamin K by epoxide reductase. This last reaction is blocked by warfarin.

proteins C and S, and other proteins found in bone (osteocalcin), kidney, spleen, pancreas, lung and other tissues [26]. Among them, the most familiar are the coagulation factors, the functions of the other proteins being less well understood [27]. The specific action of vitamin K is the posttranslational γ-carboxylation of glutamic acid residues in vitamin K-dependent proteins [28]. The process of carboxylation of proteins, in which the vitamin K cycle is involved, is illustrated in figure 1. After absorption, vitamin K is reduced to a hydroquinone (KH_2) in the presence of NADH or NADPH. The hydroquinone is then oxidized to vitamin K-2,3-epoxide (KO). This second reaction is tightly coupled with the conversion of glutamic acid residues of the proteins to γ-carboxyglutamic acid (GLA) residues in the presence of γ-carboxylase, which creates effective calcium binding sites on these proteins. Subsequently, the vitamin K epoxide is converted back to native vitamin K by epoxide reductase. Warfarin blocks this last reaction, and a congenital deficiency of the enzyme has been detected

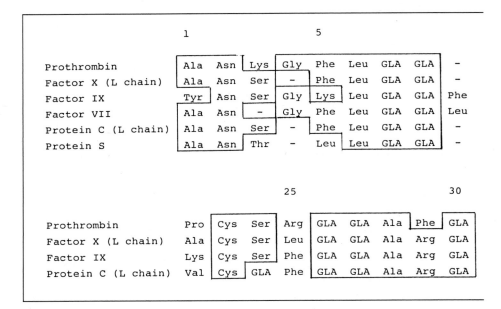

Fig. 2. Amino acid sequence of vitamin K-dependent protein (bovine). The same sequences observed at least in three of these 6 proteins are shown in boxes. Prothrombin has 10 GLA residues in its sequence.

in a clinical case very recently [29]. This vitamin K cycle appears to take place in the microsomal fraction of liver parenchymal cells.

Noncarboxylated proteins are functionally immature and can not bind calcium. Such proteins are usually found when patients have vitamin K deficiency [30] or are being treated with a vitamin K antagonistic drug, such as warfarin [31]. These proteins are generally called PIVKA, which is an abbreviation for 'proteins induced by vitamin K absence or antagonist'. For example, PIVKA-II is a non-carboxylate, immature or precursor of prothrombin (Factor II), which has 10 GLA residues in structure (fig. 2). It was accepted that the effect of warfarin on the carboxylation of prothrombin is 'all or none'. However, Esnouf and Prowse [32] and Friedman et al. [33] clearly revealed that after the administration of warfarin, prothrombins with different degrees of carboxylation are present. The same profile was observed in the case of vitamin K deficiency [20].

The formation of γ-carboxyglutamates of prothrombin and other vitamin K-dependent blood coagulation factors causes these proteins to bind

	10				15						20	
Val	Arg	Lys	Gly	Asn	Leu	GLA	Arg	GLA	Cys	Leu	GLA	GLA
Val	Lys	Gln	Gly	Asn	Leu	GLA	Arg	GLA	Cys	Leu	GLA	GLA
Val	Arg	-	Gly	Asn	Leu	GLA	Arg	GLA	Cys	Lys	GLA	GLA
Leu	-	Pro	Gly	Ser	Leu							
Leu	Arg	Pro	Gly	Asn	Val	GLA	Arg	GLA	Cys	Ser	GLA	GLA
Thr	Lys	Lys	Gly	Asn	Leu							

	35						40				
Ala	Leu	GLA	Ser	Leu	Ser	Ala	Thr	Asp	Ala	Phe	Trp
Val	Phe	GLA	Asp	Ala	GLA	Gln	Thr	Asp	GLA	Phe	Trp
Val	Phe	GLA	Asn	Thr	GLA	Lys	Thr	Thr	GLA	Phe	Trp
Ile	Phe	Gln	Asn	Thr	GLA	Asp	Thr	Met	Ala	Phe	Trp

calcium ions, which permits calcium-mediated absorption of these proteins to phospholipid membrane surfaces [34]. Thus, proteolytic (coagulation) activation potentially occurs in vivo.

Metabolism of Vitamin K

There are two types of vitamin K in nature: vitamin K_1 (phylloquinone), which is found in many plants and vegetable oil, and K_2 (menaquinone), which is synthesized by bacteria, including ones that colonize the human intestine. Both vitamin K_1 and K_2 are fat-soluble and absorbed, in the presence of bile acid, mostly in the small intestine.

The vitamin is thought to be rapidly accumulated in the liver after absorption from the intestine, while excretion occurs via the bile and, to a lesser extent, the urine [35]. The body capacity to store vitamin K is assumed to be limited [20]. A deficiency, as evidenced by hypoprothrombinemia, may arise within 24 h in animals with a biliary fistula or diversion of the intestinal lymph [20]. However, it was reported that higher molecular weight storage forms of the vitamin might exist in humans [36].

In a study on adults, on total starvation combined with antibiotic admin-istration vitamin K deficiency did not appear for 21–28 days [37]. After the ingestion of tritiated vitamin K, radioactivity persists in the plasma for 3–4 days in vitamin K-depleted man [38].

The human neonate is known to be in a precarious condition as to the vitamin K status, since the vitamin does not seem to cross the placenta easily from the mother to the fetus and vitamin K_2 synthesis by the intes-tinal flora at this time is potentially insufficient. Shearer et al. [39] reported that the cord blood vitamin K level was most probably less than one tenth the mean maternal level and was sometimes undetectable in full-term infants, even though the levels in the mothers were within normal range (0.13–0.39 ng/ml). When the mothers received vitamin K intrave-nously prior to delivery, their plasma vitamin K levels increased to between 45 and 93 ng/ml, while the levels in the cord blood of their infants ranged from undetectable to only 0.14 ng/ml [38]. Pietersma-de Bruyn and van Haard [40] found that the vitamin K_1 level in neonates was approxi-mately half that in their mothers, and Sann et al. [41] reported that in 1 of 34 infants vitamin K was not detectable at birth despite a normal level in the mother. In contrast with these early reports, Greer et al. 42] observed that the vitamin levels were similar in paired maternal and cord blood samples, although there was no significant correlation (r = 0.17) between them.

The ability of vitamin K_2 to be synthesized by the intestinal flora differs widely, depending on the strain: i.e., *Bacteroides fragilis* and some strains of *Escherichia coli* efficiently produce vitamin K_2, whereas some lactobacilli and pseudomonas species are incapable of synthesizing it [43]. The intestinal flora of breast-fed infants may produce a lower amount of vitamin K than the flora of formula-fed infants, since lactobacilli are pre-dominantly found in the former. This would be partly related to the fact that breast-fed infants are more susceptible to vitamin K deficiency in the late neonate period [18].

During the first week, the production of vitamin K by the intestinal flora seems negligible and so the diet is an important source of vitamin K in the neonate. A major K vitamer of human milk is vitamin K_1, and small amounts of menaquinones (M4–8), which might be nutritionally relevant, are also present in it [44–46]. The vitamin K_1 content range in mature milk was found to be 1.1–6.5 ng/ml by high-performance liquid chromatogra-phy (HPLC) with UV detection [47, 48], and to be 0.4–4 ng/ml [44], 0.8–4.2 ng/ml [45] and 1.6–15.2 ng/ml [47] by HPLC with fluorimetric detec-

Table 1. Laboratory tests predicting vitamin K deficiency

Increased prothrombin time
Increased partial thromboplastin time
Decreased thrombotest value or Normotest value
Decreased coagulation activity of factors II, VII, IX and X
The presence of PIVKA (protein induced by vitamin K absence)

tion, these levels being lower than originally determined on bioassaying, i.e., 15 ng/ml [49]. Similar to those of some other nutrients, the vitamin K concentration in the colostrum is apparently higher than that in mature milk [44], which would be of undoubted benefit to the newborn, since their vitamin K status is precarious, as discussed above. The concentration of vitamin K in cow milk is in the range of 4.1–18.0 ng/ml [48], and that of infant formulas is 3.8–36.7 ng/ml [31] and 13.0–33.5 ng/ml [46, 50], i.e., much higher than in mature milk.

Laboratory Tests for Evaluation of Vitamin K Status

Since measurement of the plasma vitamin K concentration involves a rather complicated procedure and requires a relatively large amount of a plasma sample [21, 38], several other convenient laboratory tests have been used for evaluation of the vitamin K status in pediatric clinics. The procedures include various means of vitamin K-dependent coagulant activity, such as measurement of the prothrombin time and partial prothrombin time, the thrombotest and Normotest (table 1). It must be noted, however, that a prolonged coagulation time or a decreased level can result from not only vitamin K deficiency but also from decreased synthesis of coagulants in the liver. Accordingly, a procedure for confirming the recovery of the coagulation time after administration of an appropriate amount of vitamin K is necessary for a diagnosis of vitamin K deficiency. Several other laboratory tests for detecting the plasma PIVKA-II level are also useful for this purpose. As discussed above, an increase in plasma PIVKA-II level indicates a state of vitamin K deficiency as long as the γ-carboxylation enzyme system is intact. Denson et al. [51] developed a method for measuring the total prothrombin activity using Taipan snake venom,

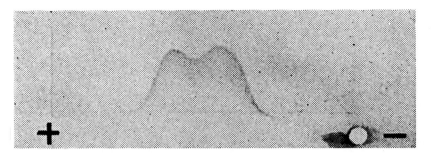

Fig. 3. Two-dimensional cross immunoelectrophoresis of prothrombin of an infant with vitamin K deficiency. The buffer contained 2 m*M* calcium lactate. Fast and slow migrated peaks indicate PIVKA-II and an active prothrombin, respectively.

which converts PIVKA-II to active prothrombin. The difference between the coagulation activities before and after the activation may indicate the PIVKA-II level in the test plasma. Another method was by Van Doorm et al. [13] and Corrigan and Kryc [52], who determined the factor II coagulant antigen level (which includes the active and nonactive forms) and coagulant activity, and then deduced the inactive prothrombin level from the data obtained. This is because active prothrombin and PIVKA-II are antigenically similar and, therefore, they cannot be determined separately by means of an ordinary immunoassay method, involving a conventional antibody against any of these proteins. In other words, an immunochemical method can be used to determine the PIVKA-II level directly only when the two forms of prothrombin can be separated by some means. For this purpose, immunoelectrophoresis has been used [53]. On cross immunoelectrophoresis, prothrombin migrates in the first dimension in a medium containing calcium. PIVKA-II will migrate further than active prothrombin. The prothrombins are then precipitated with prothrombin antibody in the second dimension. As shown in figure 3, the double peaks observed indicate two forms of prothrombin showing different calcium binding [53]. In another method involving a solution of barium carbonate, which absorbs active prothrombin, any PIVKA-II remaining after the absorption is quantitated immunologically [22]. Recently, a further improved method was developed involving a specific antibody that reacts only with PIVKA-II, i.e., not with active prothrombin [23, 54]. The latter method is known to be the most sensitive for PIVKA-II measurement and could be useful for determining the PIVKA-II concentration even in dry blood spot samples for screening vitamin K deficient infants [55]. To evaluate the vitamin K status on the basis of the plasma PIVKA-II level, it is important to note that the rate of 50% disappearance of it is approximately 50 h [56]. PIVKA-II detection can thus indicate current and also anamnestic vitamin K deficiency.

Urinary γ-carboxylated glutamic acid can also be used as a parameter for evaluating the vitamin K status, since this substance is not metabolized but is excreted as a free amino acid in the urine, but to a lesser extent with the vitamin deficiency status [57].

Vitamin K Status in Newborn Infants at Birth

As discussed above, the vitamin K concentration in cord blood was found to be approximately one tenth to half the level in the mothers [39, 41] or to be undetectable in some infants [39, 40]. Therefore on some occasions, the neonate seems to become vitamin K deficient physiologically at birth. The concentration might further decrease in the first 2–4 days of life in many infants. Although no data are available indicating daily changes in the plasma vitamin K level after birth, speculation is possible on the basis of other data on vitamin K-dependent coagulants. It is generally accepted that at birth vitamin K-dependent coagulation factors amount to 20–50% of those found in adults and the reduction is severer with increasing prematurity [58–61]. Early studies indicated that this is due to immaturity of the synthetic pathway for the factors in the infant's liver and that it is unrelated to the availability of vitamin K, because PIVKA-II in cord blood or in the neonate at birth was undetectable [13–15]. Recent studies, however, proved the presence of PIVKA-II in cord blood and in the neonate at birth, to a lesser or greater extent, ranging from 2.9 to 89% [12, 22, 53, 62–66]. The discrepancy observed by various investigators might be due to ethnic or socioeconomic differences in the populations studied, or to the sensitivity of the assay methods used. One might assume that more sensitive methods would indicate a higher prevalence rate of PIVKA-II. Radioimmunoassay or enzyme-linked immunoassay is more sensitive than crossed immunoelectrophoresis or immunophoresis, as shown by the finding that the positive PIVKA-II rates are 16.6–89% and 0–50% with the former and latter methods, respectively (table 2).

Table 2. Positive rate of PIVKA-II in cord blood

Author	Method	Positive/ studied	Positive %
Van Doorm et al. [14]	Crossed immunoelectrophoresis	0/43	0
Malia et al. [15]	Crossed immunoelectrophoresis	0/24	0
Ekelund and Hender [63]	Crossed immunoelectrophoresis	2/105	1.9
Muntean et al. [53]	Crossed immunoelectrophoresis	15/30	10.0
Sapiro et al. [12]	Immunoelectrophoresis after $BaCO_2$ absorption	27/934	2.9
Atkinson et al. [62]	Chromogenic assay with venom	48/128	38.2
Blocho et al. [64]	Clotting assay using snake venom	13/46	28
Meguro and Yamada [22]	Antibody coated beads after $BaCO_2$ absorption	2/12	16.6
Blanchard et al. [65]	RIA (hetero antibody)	161/181	89.0
Motohara et al. [66]	ELISA (monoclonal antibody)	21/99	21.2

Simultaneously obtained data, as to PIVKA-II level and clotting activity, disclosed a good correlation between abnormal levels in the coagulation test and detectable amounts of PIVKA-II in newborn infants [12, 53, 66]. In addition, it was proved that the higher the pretreatment PIVKA-II level, the greater the response, in clotting activity, to vitamin K [66].

Sann et al. [67] reported that the vitamin K_1 concentration in preterm infants (postnatal age 1–6 h) is 9 ng/ml (range 3–40 ng/ml), which is comparable to that in full-term infants at birth, determined by the same method, being 10.4 ± 5.3 ng/ml [41]. There are no other available data concerning the plasma vitamin K_1 concentration in preterm infants at birth and no evidence that the vitamin K_1 concentration is lower in preterm than in full-term infants. Thus, the much lower coagulant activity in preterm than in full-term infants may be solely attributable to the underproduction of coagulant precursors. Factor-II antigen level in full-term vs. preterm infants was found to be 44 ± 2 vs. $31 \pm 4\%$ ($100\% =$ normal adult level) by Corrigan [20], 52 ± 4 vs. 30 ± 2, 37 ± 2 and $41 \pm 3\%$ at 27–30, 31–33 and 34–36 weeks gestational age, respectively, by Sell and Corrigan [59], 51% (range 29–76%) vs. 31% (range 19–54%) by Barnard et al. [60], and 66.1 ± 6.4 vs. $35 \pm 9.4\%$ by Ogata et al. [61]. Reflecting the lower level of factor-II antigen in premature infants, the prevalence rate of positive PIVKA-II in their plasma (2.8%) was apparently lower than that in full-term infants during the first 24 h (52.9%) [66]. As to coagulation activity, only 9% of preterm infants with reduced normotest results, of less than 30%, exhibited detectable PIVKA-II levels at 1 day [61]. These observations may well explain the fact that a beneficial effect of vitamin K on the levels of the vitamin K-dependent factors was not observed in 31 preterm infants [17]. After development of the factor-II synthesis pathway during the first several days, the positive rate of plasma PIVKA-II in preterm infants increased to a level equivalent to that in full term infants at birth, when vitamin K was not given to them [61].

Dam et al. [3] noted that infants who suffered from intrauterine asphyxia had lower prothrombin levels than infants without this complication. Corrigan and Kryc [52] found that newborns with low Apgar scores and meconium staining had lower factor-II levels and a lower ratio of coagulant activity to antigen, compared with infants with high Apgar scores, indicating a possible increase in the PIVKA-II level. Conversely, other recent studies disclosed that the plasma PIVKA-II level was not different between infants with high and low Apgar scores, as determined with a direct measurement method [12, 61]. The development of the enzymes involved in the vitamin K cycle, such as γ-carboxylase, in the fetus should be investigated in the future, since there are no data concerning this issue at present.

Hemorrhagic Disease of the Newborn

Lane and Hathaway [18] classified cases as to the pattern of vitamin K deficiency hemorrhage occurring in infancy into three groups, early hemorrhagic disease of the newborn (HDN), classical HDN and late neonatal hemorrhagic disease (late neonatal vitamin K deficiency).

Early Hemorrhagic Disease of the Newborn

Moderate to severe hemorrhage due to vitamin K deficiency observed at the time of delivery or during the first 24 h is regarded as early HDN. The extent of bleeding varies from umbilical bleeding or skin bruising to giant cephalohematomas, gastrointestinal bleeding, and widespread and fatal intracranial, intrathoracic, and intraabdominal hemorrhage [68–74]. Although some idiopathic cases have been reported [68, 69], the disease is mostly seen in infants whose mothers have received drugs that affect vitamin K metabolism. These drugs include anticonvulsants [70–74], warfarin [75], rifampin and isoniazid [76].

Maternal anticonvulsants taken during pregnancy increase the incidence of early HDN [70–74]. The coagulation defect is thought to be similar to that seen in the vitamin K deficient status [70] and the incidence of HDN has been lowered by antenatal vitamin K administration [71]. Prolongation of the prothrombin time and an increased rate of positive PIVKA-II have been observed in adults given anticonvulsants [77, 78]. Most cases of HDN induced by maternal anticonvulsants have involved barbiturates or phenytoin, or both. However, Davies et al. [78] suggested that other anticonvulsants, such as carbamazepine, valproate and clonamzepam, cannot be excluded from this category. The defect may not depend on the particular anticonvulsant used. Thus, routine antenatal vitamin K administration seems to be indicated in pregnancies complicated by anticonvulsants, although we need more data concerning the exact risk for this complication in infants born to epileptic mothers with different nutritional status.

Therapeutic doses of warfarin during pregnancy may result in early HDN [75]. In other cases born to mothers taking warfarin, several phenotypic abnormalities were recorded, such as irregular ossification, nasal hypoplasia, distal digital hypoplasia and mild conductive hearing loss, categorized as 'warfarin embryopathy'. The first two characteristics are uniformly present in such infants, whereas the others are more variable but consistent with the disorder [79]. Recently, an inborn deficiency of vitamin K epoxide reductase (warfarin is an inhibitor of this enzyme) was found in a boy who had excessive bleeding and bruising from birth [29]. His clinical features were quite comparable to those of warfarin embryopathy. Similar to the cases of other vitamin K-dependent proteins, the γ-carboxylation of osteocalcin allows calcium binding and the modified osteocalcin may play an important role in early calcium deposition in bone development [80]. Warfarin's pharmacologic inhibition of the γ-carboxylation of osteocalcin may result in abnormalities of calcium deposition (slipped epiphyses) and bone development (nasal distal hypoplasia). Therefore, the disorder of vitamin K epoxide reductase might be understood as a 'genocopy' of warfarin embryopathy.

Infants born to mothers taking rifampicin and isoniazid during pregnancy may be at risk for early HDN. In six such pregnancies, three of the infants developed the disorder, with one infant death and one maternal death caused by hemorrhage [76].

Classical Hemorrhagic Disease of the Newborn (HDN)

Classical HDN occurs from 1 to 7 days, mostly from 2 to 5 days of age [18, 20, 74, 81]. The affected infants are normal at birth and then develop gastrointestinal bleeding, generalized echimosis, nasal bleeding or bleeding following circumcision. Intracranial bleeding, which is a characteristic feature in late neonatal vitamin K deficiency, is less common at this age [18, 20]. Immunologic and coagulation techniques have revealed the same laboratory patterns in affected newborns as those in older children and adults with vitamin K deficiency. That is, these infants have a low level of activated prothrombin and a higher level of the prothrombin precursor protein [53, 54, 82].

Classical HDN occurs mainly in breast-fed infants, and rarely in infants receiving supplementary or exclusive formulae, or in infants with prophylactic administration of vitamin K at birth [20, 43, 81]. The development of the disorder appears to be related to the total (dietary) intake of vitamin K, since bacterial synthesis of the vitamin in the intestine is negligible in infants at this stage. The factors responsible for this might be as follows: (i) whether the infants received vitamin K prophylactically at birth; (ii) what type of feeding, such as breast, formula or mixed, they received; (iii) if they were fed breast milk, how much vitamin K was contained in it; and (iv) what was the total milk ingestion, regardless of the type of feeding.

The vitamin K concentration in breast milk may vary widely, reflecting the maternal status of the vitamin. For instance, when vitamin K (10 mg/day) was given to mothers, the vitamin concentration in milk increased to more than 50 ng/ml, which is approximately 2 times the concentration in infant formulae [83]. However, without the administration of vitamin K, the concentration in human milk is lower than that in infant formulae during the first week of lactation [44, 46]. The lower level of vitamin K in breast milk reflects the level of carboxyprothrombin in the infants. When breast-fed infants were not given vitamin K at birth, the PIVKA-II positive prevalence rate increased in cord blood from 22.9 to 61.5 and 52.9% in infants at 3 and 5 days of age, respectively [66]. Another study also revealed that the PIVKA-II positive rate increased in cord blood from 0 to more than 50% in breast-fed infants during the first 5 days of life [84]. The infants who are given milk formulae showed a much lower incidence of PIVKA-II positivity at 5 days of age.

The amount of milk ingested is also a factor in the development of vitamin K deficiency during this period. Kries et al. [85] proposed that the type of milk feeding, breast or formula, might be less important for a baby's vitamin K supply than the actual milk intake. Motohara et al. [86] evaluated the relationship between the milk intake and the extent of the vitamin K deficiency. They found that the average total milk intake during the first 3 days of life was significantly lower in PIVKA-II positive infants than in PIVKA-II negative infants, both with breast and mixed (breast and formula) feeding. In addition, there was a significantly negative correlation between the PIVKA-II positive proportion and the milk intake in breast-fed infants. A highly elevated PIVKA-II level, indicating a high risk for classical HDN, was found in 3 of 257 breast-fed infants in their study [86]. They reported that the total amounts of milk ingested in the first 3 days in these 3 infants were 25, 52 and 66 ml, respectively (more than 300 ml is required for maintaining a sufficient vitamin K status) [86]. Thus, a low intake of maternal milk or milk formulae during the first several days of life is more relevant as to vitamin K deficiency in this period.

Some unknown cause of vitamin K malabsorption may be another factor for classical HDN. It was evident in some infants that the oral administration of vitamin K did not result in a sufficient increase in the plasma vitamin K level [87]. The appearance of PIVKA-II in the plasma of some infants who received oral vitamin K at birth may be related to mal-absorption of the vitamin K [86].

Late Neonatal Vitamin K Deficiency

Vitamin K deficiency can occur in older infants due to an unknown cause (idiopathic) or with secondary manifestation of underlying disorders. The former type occurs at 2–8 weeks of age, whereas the latter type may be seen any time during the first year [18, 88]. The underlying diseases for the secondary vitamin K deficiency are listed in table 3. The diseases mostly involve impaired intestinal absorption of the vitamin, examples being prolonged diarrhea, congenital bile duct atresia, hepatitis, α_1-anti-trypsin deficiency and celiac disease [18, 89, 91]. Decreased vitamin K_2 synthesis by the intestinal flora secondary to the use of antibiotics [92] and prolonged exposure to warfarin-contaminated talcum powder [93] were other causes of secondary vitamin K deficiency. In most cases of late neo-natal vitamin K deficiency, however, no underlying diseases have been identified. More than 500 cases of idiopathic vitamin K deficiency of the late neonatal period have been reported during the last 10 years from many countries: Taiwan [94], Singapore [95, 96], Thailand [97, 98], Japan [45, 99, 100], England [74, 101–103], the United States [104–106], Germany

Table 3. Late onset vitamin K deficiency underlying disease or factors

α_1-Antitrypsin deficiency	Chronic diarrhea
Abetalipoproteinaemia	Chronic warfarin exposure
Antibiotics (cephalosporins)	Cystic fibrosis
Bile duct atresia	Subclinical choleostasis
Celiac disease	

[88, 107], the Netherlands [108] and so on, among which Asian countries appear to have a higher incidence than Western countries. The disorder has been recorded in 1 out of 4,500 babies in Japan [100] and the ratio has been estimated to be 1 out of 50,000–100,000 babies in Germany [88].

The initial symptoms of the disorder are vomiting, lethargy, hemorrhage, convulsions, pallor, loss of appetite, fever, unconsciousness, dyspnea and others [18, 94, 100]. A main, and the most unfavorable, clinical feature is acute intracranial hemorrhage, which was observed in 50–80% of the affected infants [88, 94, 100]. The second common clinical feature is so-called nodular purpura, widespread deep echymosis. Gastrointestinal or mucosal membrane bleeding is also an initial symptom, although it is uncommon [18]. The outcome for the affected infants with intracranial bleeding is death or a severely handicapped status in 50–70% of these infants [45, 88, 90, 94].

It must be noted that most of the infants with idiopathic vitamin K deficiency and many of those with putative secondary deficiency did not receive vitamin K at birth and were exclusively breast-fed [18, 88, 100]. A nationwide survey in Japan indicated that 90.6 and 66.6% of the affected infants fed breast milk exclusively had idiopathic and secondary late neonate vitamin K deficiency, respectively [100]. Kries et al. [88] reviewed numerous reported cases of the disorder, of which 186 out of 198 had been exclusively breast-fed, whereas only 3 had been exclusively formula-fed. A similar observation was reported for cases of subclinical vitamin K deficiency, which was identified in a screening test [55].

As mentioned above, most of the infants with late neonatal vitamin K deficiency did not undergo vitamin K prophylaxis at birth. However, this treatment does not completely prevent the bleeding. Verity et al. [109] reported three infants with the disorder who had received 1 mg of vitamin K at birth. Motohara et al. [55] noticed that a highly elevated PIVKA-II level indicating severe vitamin K deficiency was detected in some infants receiving vitamin K at birth. When the half-life of vitamin K in infants is taken into account, the administered vitamin K may have little influence on the deficiency developing in older neonates [38, 110]. Another factor is the vitamin K_1 content of breast milk. The concentration of vitamin K_1, the main form of vitamin K in human milk, is highest in the colostrum and decreases gradually with lactation, and furthermore, the concentration varies widely [44]. Approximately 1 month after obstetrical delivery, human milk contains roughly 1/3–1/20 of the vitamin K_1 present in commercial formulae in Japan [46, 50], as well as in other countries [48]. Some authors suggested that the lower content of vitamin K in breast milk related to the late neonatal vitamin K deficiency [18, 94, 105, 106]. However, the vitamin K_1 content of breast milk obtained from mothers who had affected infants was found to be lower in only some and not in most cases, when compared with milk samples from mothers of unaffected babies [44–46, 111]. Therefore, there seem to be other causes for the vitamin K deficiency in breast-fed infants. The bacterial content in the intestine may be one of the possible factors in the late neonatal vitamin K deficiency.

It has been disclosed by several studies that mild liver dysfunction is often a feature of patients considered to have idiopathic vitamin K deficiency [45, 94, 97]. Liver dysfunction to a similar degree was also seen in cases of vitamin K deficiency associated with α_1 antitrypsin deficiency or neonatal hepatitis [88]. Matsuda et al. [111] studied infants with mild to severe vitamin K deficiency, which was detected in a screening test involving the measurement of PIVKA-II in dry blood spots at one month of age. These babies showed normal levels of SGOT and SGPT, which coincided with early criteria of idiopathic vitamin K deficiency. However, they had higher serum concentrations of bile acid and alkaline phosphatase, accompanied by a slight elevation of serum direct bilirubin, and a decreased serum concentration of 25-hydroxy vitamin D. They were all breast-fed and the vitamin K_1 contents in the milk were all within the control range. The data indicated that the babies had subclinical cholestasis, as a cause of 'idiopathic' vitamin K deficiency. Thus, it is speculated that the late neonatal vitamin K deficiency predominantly occurring in breast-fed infants is not solely due to a lower vitamin K content but rather to a combination with other factors including subclinical liver dysfunction. The higher contents of vitamin K in the formulas [46, 48, 50] could mask the deficiency and, on the contrary, the lower content in breast milk may accelerate the development of a deficiency in the vitamin.

Vitamin K Prophylaxis in Infants

The National Academy of Science recommended in 1980 that the safe and adequate dietary intake of vitamin K for infants is 12 µg/day [112]. Recently, Motohara et al. [86] reported that the minimum dose of vitamin K required to prevent a positive PIVKA-II result was 15 µg in babies on the basis of the results of an oral vitamin K supplementation study. For the ingestion of this much vitamin K, a total of more than 500 ml of breast milk should be given to infants during the first 3 days [86]. There are many infants who receive much lower amounts of breast milk and some infants have malabsorption of vitamin K. The prophylactic administration of vitamin K seems to be the most reasonable way to prevent HDN. The oral administration of 1 mg of vitamin K with the first feed reduced the rate of detection of plasma PIVKA-II on day 5 from 48% to 0, whereas the detection rate on day 5 was about 50% in babies not given vitamin K [113]. Dunn [114] observed no cases of classical hemorrhagic disease in neonate (HDN) among 30,000 newborns given 1 mg of oral vitamin K at birth. Thus, the recommendation of the Committee on Nutrition of the American Academy of Pediatrics, i.e., 0.5–1.0 mg of vitamin K parentally [9], seems rational to prevent vitamin K deficiency in the neonate, including premature babies [61].

Prevention of the late neonatal hemorrhagic disease due to vitamin K deficiency is another clinical problem. All available data indicate that most of the affected infants were exclusively breast-fed and did not receive vitamin K at birth [18, 88, 100]. There have been few prospective studies. Motohara et al. [115] studied the effect of oral administration of vitamin K on PIVKA-II detection rate in breast-fed infants. At 1 month of age, no significant difference of PIVKA-II positive rate was observed between infants receiving 5 mg vitamin K once at birth and those receiving no vitamin K prophylaxis. The observation was subsequently confirmed in a screening study on 19,029 infants [55]. However, when the study was performed in a larger scale including 30,704 infants, it was evident that the PIVKA-II detection rate, including severe and mild deficiency of the vitamin, was significantly lower in infant groups receiving vitamin K at birth (2 mg) and two times 2 mg at birth and 1 week of age, respectively, compared with those in infants without prophylaxis. The infants receiving vitamin K two times showed the lowest proportion of PIVKA-II [116] (table 4). On the basis of the fact that some infants with vitamin K deficiency received vitamin K two or three times since birth [100] and that the

Table 4. PIVKA-II positive infants according to method of feeding and vitamin K prophylaxis at birth

Vitamin K prophylaxis	Group	Number	PIVKA-II > 4.0 AU/ml[a]		PIVKA-II > 20 AU/ml[a]	
			solely breast-fed	mixed or formula-fed	solely breast-fed	mixed or formula-fed
None	G1	11,726 (6,993)[b]	42 (0.60%)	13 (0.27%)	11 (0.16%)	1 (0.02%)
2 mg orally on day 1	G2	10,242 (5,392)[b]	14 (0.26%)	5 (0.1%)	2 (0.04%)	0
2 mg orally on day 1 and 2–4 mg on day 7	G3	8,736 (5,188)[b]	7 (0.13%)	0	1 (0.02%)	0
Total		30,704	63 (0.36%)	18 (0.14%)	14 (0.08%)	1 (0.01%)

p < 0.05 (G1 vs G2, PIVKA-II > 4.0, solely breast-fed)
p < 0.05 (G1 solely breast-fed vs mixed or formula-fed)
p < 0.05 (G1 vs G2)
p < 0.005 (G2 vs G3)
p < 0.05 (Total, PIVKA-II > 4.0)
p < 0.05 (Total, PIVKA-II > 20)

Statistical significance of the vitamin K deficiency in each group was compared using the χ^2 test with Yet's correction.

[a] AU = Arbitrary unit.

[b] Number of solely breast-fed infants.

biological half-life of vitamin K was estimated to be less than 72 h [38], routine multiple administration of vitamin K, such as at birth, at 1 week and at 3–4 weeks of age, seems to be necessary to prevent vitamin K deficiency nearly completely, especially in solely breast-fed infants.

References

1 Townsend, C.W.: The hemorrhagic disease of the newborn. Archs Pediat *11:* 559 (1894).

2 Dam, H.: The antihemorrhagic vitamin in the chick. Biochem. J. *29:* 1273 (1935).

3 Dam, H.; Dyggve, H.; Larsen, H.; Plum, D.: The relation of vitamin K deficiency to hemorrhagic disease of the newborn. Adv. Pediat. *5:* 129 (1952).

4   Brinkhouse, K.M.; Smith, H.D.; Warner, E.D.: Plasma prothrombin level in normal infancy and in hemorrhagic disease of the newborn. Am. J. Med. Sci. *193:* 475 (1937).

5   Waddell, W.W.; Guerry, D.: The role of vitamin K in the etiology, prevention, and treatment of hemorrhage in the newborn infant. J. Pediat. *15:* 802 (1939).

6   Nggaard, K.K.: Prophylactic and curative effect of vitamin K in hemorrhagic disease of the newborn. Acta obstet. gynec. scand. *19:* 361 (1939).

7   Wefring K.W.: Hemorrhage in the newborn and vitamin K prophylaxis. J. Pediat. *61:* 686 (1962).

8   Aballi, A.J.; de Lamerens, S.: Coagulation changes in the neonatal period and early infancy. Pediat. Clins N. Am. *9:* 785 (1962).

9   Committee on Nutrition, American Academy of Pediatrics: Vitamin K compounds and the water-soluble analogues: use in therapy and prophylaxis in pediatrics. Pediatrics *28:* 501 (1961).

10 Sutherland, J.M.; Glueck, H.I.; Gleser, G.: Hemorrhagic disease of the newborn. Am. J. Dis. Child. *113:* 524 (1967).

11 Keenan, W.J.; Jewett, T.; Glueck, H.I.: Role of feeding and vitamin K in hypoprothrombinemia in the newborn. Am. J. Dis. Child. *121:* 271 (1971).

12 Shapiro, A.D.; Jacobson, L.J.; Armon, M.E.; Manco-Johnson, M.J.; Hulac, P.; Lane, P.; Hathaway, W.E.: Vitamin K deficiency in the newborn infant: prevalence and perinatal risk factors. J. Pediat. *109:* 675 (1986).

13 Van Doorm, J.M.; Muller, A.D.; Hemker, H.C.: Heparin-like inhibitor, not vitamin K deficiency in the newborn. Lancet *i:* 852 (1977).

14 Van Doorm, J.M.; Hemker, H.C.: Vitamin K deficiency in the newborn. Lancet *ii:* 708 (1977).

15 Malia, R.G.; Preston, F.E.; Mitchell, V.E.: Evidence against vitamin K deficiency in normal neonate. Thromb Haemostasis *44:* 159 (1980).

16 Gobel, U.; Sonnenschein-Kosenow, S.; Petrich, C.; Voss, H.V.: Vitamin K deficiency in the newborn. Lancet *ii:* 187 (1977).

17 Mori, P.G.; Biogni, C.; Odino, S.; Tonini, G.D.; Boeri, E.; Serra, G.; Romano, C.: Vitamin K deficiency in the newborn. Lancet *ii:* 186 (1977).

18 Lane, P.A.; Hathaway, W.E.: Vitamin K in infancy. J. Pediat. *106:* 351 (1985).

19 Goldman, H.I.; Deposito, F.: Hypoprothrombinemic bleeding in young infants. Am. J. Dis. Child. *111:* 430 (1966).

20 Corrigan, J.: Vitamin K-dependent proteins. Adv. Pediat. *28:* 57 (1981).

21 Shearer, M.J.: High-performance liquid chromatography of K vitamins and their antagonists; in Giddings, Grushka, Cazes, Brown, Advances in chromatography, p. 243 (Dekker, New York 1983).

22 Meguro, M.; Yamada, K.: A simple and rapid test for PIVKA-II in plasma. Thromb. Res. *25:* 109 (1982).

23 Motohara, K.; Kuroki, Y.; Kan, H.; Endo, F.; Matsuda, I.: Detection of vitamin K deficiency by use of an enzyme-linked immunosorbent assay for circulating abnormal prothrombin. Pediat. Res. *19:* 354 (1985).

24 Forbes, D.: Delayed presentation of hemorrhagic disease of the newborn. Med. J. Aust. *ii:* 136 (1983).

25 Lane, P.A.; Hathaway, W.E.; Githens, J.H.; Krugman, P.D.; Rosenberg, D.A.: Fatal intracranial hemorrhage in a normal infant secondary to vitamin K deficiency. Pediatrics *72:* 562 (1983).

26   Hauschka, P.V.; Lian, J.B.; Gallop, P.M.: Vitamin K and mineralization. Trends
      Biochem. Sci. *3:* 375 (1978).
27   Gallop, P.M.; Lian, J.B.; Hauschka, P.V.: Carboxylated calcium binding protein and
      vitamin K. New Engl. J. Med. *302:* 1460 (1980).
28   Stenflo, J.; Fernlund, P.; Egan, W.: Vitamin K-dependent modifications of glutamic
      acid residues in prothrombin. Proc. natn. Acad. Sci. USA *71:* 2730 (1974).
29   Pauli, R.M.; Lian, J.B.; Mosher, D.F.; Stuffie, J.W.: Association of congenital defi-
      ciency of multiple vitamin K-dependent coagulation factors and the phenotype of
      the warfarin embryopathy: clues to the mechanism of teratology of coumarin deriv-
      atives. Am. J. hum. Genet. *41:* 566 (1987).
30   Hemker, H.C.; Muller, A.D.; Loelinger, E.A.: Two types of prothrombin in vitamin
      K deficiency. Thromb. Diath. haemorrh. *23:* 633 (1970).
31   Ganrot, P.D.; Nilehn, J.E.: Plasma prothrombin during treatment with dicumarol.
      II. Demonstration of an abnormal prothrombin fraction. Scand. J. Lab. Invest. *22:*
      23 (1968).
32   Esnouf, M.D.; Prowse, C.V.: The gamma-carboxy glutamic acid content of human
      and bovine prothrombin following warfarin treatment. Biochim. biophys. Acta *490:*
      471 (1977).
33   Friedman, P.A.; Rosenberg, R.D.; Haushka, D.V.; Fitz-James, A.: A spectrum of
      partially carboxylated prothrombin in the plasma of warfarin-treated patients. Bio-
      chim. biophys. Acta *494:* 271 (1977).
34   Suttie, J.W.; Jackson, C.M.: Prothrombin structure, activation, and biosynthesis.
      Physiol. Rev. *57:* 1 (1977).
35   Shearer, M.J.; Mallinson, C.N.; Webster, G.R.; Barkham, P.: Clearance from plasma
      and excretion in urine, feces and bile of an intravenous dose of tritiated vitamin K_1
      in man. Br. J. Haemat. *22:* 579 (1972).
36   Bjornsson, T.D.; Meffin, P.J.; Swezey, S.E.: Disposition and turnover of vitamin K_1
      in man; in Suttie, Vitamin K metabolism and vitamin K-dependent proteins, p. 328
      (University Park Press, Baltimore 1980).
37   Frick, P.G.; Riedler, G.; Biogli, H.: Dose response and minimal daily requirement
      for vitamin K in man. J. appl. Physiol. *23:* 387 (1969).
38   Shearer, M.J.; Barkan, D.K.; Webster, G.R.: Absorption and excretion of an oral
      dose of tritiated vitamin K_1 in man. Br. J. Haemat. *18:* 297 (1970).
39   Shearer, M.J.; Rahim, S.; Barkhan, P.; Stimmler, L.: Plasma vitamin K_1 in mothers
      and their newborn babies. Lancet *ii:* 460 (1982).
40   Pietersma-de Bruyn, A.L.J.M.; Haard, P.M.M. van: Vitamin K_1 in the newborn.
      Clin. Chim. Acta *150:* 95 (1985).
41   Sann, L.; Leclercq, M.; Troncy, J.; Guillamound, M.; Berland, M.; Coeur, P.: Serum
      vitamin K_1 concentration and vitamin K-dependent clotting factor activity in
      maternal and fetal cord blood. Am. J. Obstet. Gynec. *153:* 771 (1985).
42   Greer, F.R.; Mummah-Schendel, L.L.; Marshall, S.; Suttie, J.W.: Vitamin K_1 (phyl-
      loquinone) and vitamin K_2 (menaquinone) status in newborns during the first week
      of life. Pediatrics *81:* 137 (1988).
43   Bently, R.; Meganathan, R.: Biosynthesis of vitamin K_2 (menaquinone) in bacteria.
      Microbiol. Rev. *46:* 241 (1982).
44   Kries, R.V.; Shearer, M.; Mccarrthy, P.T.; Hang, M.; Harzer, G.; Gobel, U.: Vitamin

K₁ content of maternal milk: influence of the stage of lactation, lipid composition, and vitamin K_1 supplements given to the mother. Pediat. Res. *22:* 513 (1987).

45 Motohara, K.; Matsukura, M.; Matsuda, I.; Iribe, K.; Ikeda, T.; Kondo, Y.; Yonekubo, A.; Yamamoto, Y.; Tsuchiya, F.: Severe vitamin K deficiency in breast-fed infants. J. Pediat. *105:* 943 (1984).

46 Shirahata, A.; Nojiri, T.; Miyaji, Y.; Yamada, K.; Ikeda, I.; Takii, I.; Suzuki, E.; Sato, S.: Vitamin K contents of infant formula products and breast milk. Igaku No Ayumi *118:* 857 (1981) (in Japanese).

47 Haroon, Y.; Shearer, M.J.; Gun, W.G.; McEnery, G.; Barkhan, P.: The content of phylloquinone (vitamin K_1) determined by high-performance liquid chromatography. J. Nutr. *112:* 1105 (1982).

48 Shearer, M.J.; Allan, N.; Harron, V.; Barkhan, P.: Nutritional aspects of vitamin K in the human; in Suttie, Vitamin K metabolism and vitamin K-dependent proteins, p. 317 (University Park Press, Baltimore 1979).

49 Goldman, H.I.; Amadio, P.: Vitamin K deficiency after the newborn period. Pediatrics *44:* 745 (1969).

50 Tsuchiya, F.: Vitamin K contents in commercial formulas in Japan. Perinat. Med. *12:* 1125 (1982) (in Japanese).

51 Denson, K.W.E.; Barrett, R.; Briggs, R.: The specific assay of prothrombin using the Taipan snake venom. Br. J. Haemat. *21:* 219 (1971).

52 Corrigan, J.J., Jr.; Kryc, J.J.: Factor II (prothrombin) levels in cord blood: correlation of coagulant activity with immunoreactive protein. J. Pediat. *97:* 979 (1980).

53 Muntean, W.; Petek, W.; Rosanelli, K.; Mutz, I.D.: Immunologic studies of prothrombin in newborns. Pediat. Res. *13:* 1262 (1979).

54 Blanchard, R.A.; Furie, B.C.; Kruger, S.F.; Furie, B.: Acquired vitamin K-dependent carboxylation deficiency in liver disease. New Engl. J. Med. *305:* 242 (1981).

55 Motohara, K.; Endo, F.; Matsuda, I.: Screening for late neonatal vitamin K deficiency by acarboxyprothrombin in dried blood spots. Archs Dis. Childh. *62:* 370 (1987).

56 Shapiro, A.D.; Hulac, P.; Jacobson, L.J.; Lane, P.A.; Manco-Johnson, M.J.; Hathaway, W.E.: Prevalence of vitamin K deficiency in newborn infants. Thromb. Haemostasis *54:* 125 (1985).

57 Sann, L.; Leclercq, M.; Fouillit, M.; Chapuis, M.C.; Bruyère, A.: Gamma carboxyglutamic acid in urine of newborn infants. Clin. Chim. Acta *142:* 32 (1984).

58 Jensen, A.H.B.; Josso, F.; Zamet, D.; Moneset-Couchard, M.; Minkowski, A.: Evolution of blood clotting factor levels in premature infants during the first ten days of life: a study of 96 cases with comparison between clinical status and blood clotting factor levels. Pediat. Res. *7:* 638 (1973).

59 Sell, E.J.; Corrigan, J.J.: Platelet counts, fibrinogen concentrations and factor V and factor VIII levels in healthy infants according to gestational age. J. Pediat. *82:* 1028 (1973).

60 Barnard, D.R.; Simmons, M.A.; Hathaway, W.E.: Coagulation studies in extremely premature infants. Pediat. Res. *13:* 1330 (1979).

61 Ogata, T.; Motohara, K.; Endo, F.; Kondo, Y.; Ikeda, T.; Kudo, Y.; Iribe, K.; Matsuda, I.: Vitamin K effect in low birth weight infants. Pediatrics *81:* 423 (1988).

62 Atkinson, P.M.; Bradlow, B.A.; Monlinean, J.D.: Acarboxy-prothrombin in cord plasma from normal neonates. J. Pediat. Gastroenterol. Nutr. *3:* 450 (1984).

63   Ekelund, H.; Hender, U.: Prothrombin and vitamin K deficiency in the newborn. Eur. Paediat. Haematol. Oncol. *1:* 59 (1984).

64   Bloch, C.A.; Rothberg, A.D.; Bradlow, B.A.: Mother-infant prothrombin precursor status at birth. J. Pediat. Gastroenterol. Nutr. *3:* 101 (1984).

65   Blanchard, R.A.; Furie, B.C.; Barnett, J.: Vitamin K deficiency in newborns and their mothers. Thromb. Haemostasis *54:* 1340 (1985).

66   Motohara, K.; Endo, F.; Matsuda, I.: Effect of vitamin K administration on acarboxy-prothrombin (PIVKA-II) levels in newborns. Lancet *ii:* 242 (1985).

67   Sann, L.; Leclercq, M.; Guillamount, M.; Trouyez, R.; Bethend, M.; Bourgeay-Causse, M.: Serum vitamin K_1 concentration after oral administration of vitamin K_1 in low birth weight infants. J. Pediat. *107:* 608 (1985).

68   Leonard, S.; Anthony, B.: Giant cephalohematoma of newborn with hemorrhagic disease and hyperbilirubinemia. Am. J. Dis. Child. *101:* 170 (1961).

69   Hull, M.G.; Wilson, J.A.: Massive scalp hemorrhage after fetal blood sampling due to hemorrhagic disease. Br. Med. J. *iv:* 321 (1972).

70   Mountain, K.R.; Mirsh, J.; Gall, A.S.: Neonatal coagulation defect due to anticonvulsant drug treatment in pregnancy. Lancet *i:* 265 (1970).

71   Davis, P.P.: Coagulation defect due to anticonvulsant drug treatment in pregnancy. Lancet *i:* 413 (1970).

72   Bleyer, W.A.; Skinner, A.L.: Fatal neonatal haemorrhage after maternal anticonvulsant therapy. J. Am. med. Ass. *235:* 626 (1976).

73   Deblay, M.F.; Vert, P.; Andre, M.; Marchal, F.: Transplacental vitamin K prevents haemorrhagic disease of the infants of epileptic mothers. Lancet *i:* 1247 (1982).

74   McNinch, A.W.; Orme, R.L.E.; Tripp, J.H.: Haemorrhagic disease of the newborn returns. Lancet *i:* 1089 (1983).

75   Stevenson, R.E.; Burton, U.M.; Ferlanto, G.J.; Hazards of oral anticoagulants during pregnancy. J. Am. med. Ass. *235:* 626 (1976).

76   Eggermont, E.; Logghe, M.; Casseye, W. van de: Haemorrhagic disease of the newborn in the offspring of rifampin and isoniazid treated mothers. Acta paediat. belg. *29:* 87 (1976).

77   Andreasen, P.B.; Lyngbye, J.; Trolle, E.: Abnormalities in liver function tests during long-term diphenylhydantoin therapy in epileptic outpatients. Acta med. scand. *194:* 261 (1973).

78   Davies, V.; Rothberg, A.D.; Argent, A.C.; Atkinson, P.M.; Staub, H.; Pienaar, N.L.: Precursor prothrombin status in patients receiving anticonvulsant drugs. Lancet *i:* 126 (1985).

79   Hall, J.G.; Pauli, R.M.; Wilson, K.M.: Maternal and fetal sequelae of anticoagulation during pregnancy. Am. J. Med. *68:* 122 (1980).

80   Hauschka, P.V.; Reid, M.L.: Timed appearance of a calcium-binding protein containing γ-carboxyglutamatic acid in developing chick bone. Devl. Biol. *65:* 426 (1978).

81   Sutherland, J.M.; Glueck, H.I.; Gleser, G.: Hemorrhagic disease of the newborn: breast feeding as a necessary factor in the pathogenesis. Am. J. Dis. Child. *113:* 524 (1967).

82   Dreyfus, M.; Leolong-Tisser, M.C.; Lonbard, L.; Tchernia, G.: Vitamin K deficiency in the newborn. Lancet *i:* 1351 (1979).

83   Umezawa, S.; Hanawa, Y.; Yamamoto, Y.; Yonekubo, A.: The vitamin K_2 content

in breast milk after maternal ingestion. Igaku No Ayumi *135:* 1105 (1985) (in Japanese).

84 Kries, R.V.; Gobel, U.; Masse, B.: Vitamin K deficiency in the newborn. Lancet *ii:* 728 (1985).

85 Kries, R.V.; Becker, A.; Gobel, U.: Vitamin K in the newborn: influence of nutritional factors on acarboxy-prothrombin detectability and factor II and VII clotting activity. Eur. J. Pediat. *146:* 123 (1987).

86 Motohara, K.; Matsukane, I.; Endo, F.; Kiyota, Y.; Matsuda, I.: Relationship of milk intake and vitamin K supplementation to vitamin K status in newborn. Pediatrics *84:* 90 (1989).

87 McNinch, A.W.; Upton, C.; Samuels, M.; Shearer, M.J.; McCarthy, P.; Tripp, J.H.; Orme, R.L.E.: Plasma concentration after oral or intramuscular vitamin K_1 in neonates. Archs Dis. Childh. *60:* 814 (1985).

88 Kries, R.V.; Shearer, M.J.; Gobel, U.: Vitamin K in infancy. Eur. J. Pediat. *147:* 106 (1988).

89 Hope, P.L.; Hall, M.A.; Hillward-Sadler, P.: Alpha 1-antitrypsin deficiency presenting as a bleeding diathesis in the newborn. Archs Dis. Childh. *57:* 68 (1982).

90 Caballerd, F.M.; Buchana, G.R.: Abetalipoproteinemia presenting as severe vitamin K deficiency. Pediatrics *65:* 161 (1980).

91 Poley, J.R.; Hamphrey, G.B.: Bleeding disorder in an infant associated with anicteric hepatitis. Clin. Pediat. *13:* 1045 (1974).

92 Chan-Lui, W.Y.; Stoebel, B.A.; Young, C.Y.: Hemorrhagic tendency as a complication of moxalactum therapy in bacterial meningitis. Brain Dev. *5:* 417 (1983).

93 Martin-Bouyer, G.; Khanh, N.B.; Linh, P.D.; Hoa, D.Q.; Tuan, L.C.; Tournean, J.; Barin, C.; Guerbois, H.; Binh, T.V.: Epidemic of hemorrhagic disease in Vietnamese infants caused by warfarin-containing talcs. Lancet *i:* 230 (1983).

94 Chaou, W.T.; Chou, M.L.; Etzman, D.V.: Intracranial hemorrhage and vitamin K deficiency in early infancy. J. Pediat. *105:* 880 (1984).

95 Chan, M.C.K.; Wong, H.B.: Late hemorrhagic disease of Singapore infants. J. Singapore Paediat. Soc. *9:* 72 (1967).

96 Hon, T.E.: Severe hypoprothrombinaemic bleeding in the breast fed young infant. Singapore Med. J. *10:* 43 (1969).

97 Bhanchet, P.; Uchinda, S.; Hathirat, P.; Visudhiphan, P.; Bhamaraphavati, N.; Bukkavesa, S.: A bleeding syndrome in infants due to acquired prothrombin complex deficiency: a survey of 93 affected infants. Clin. Pediat. *16:* 992 (1977).

98 Mitrakul, C.; Tinakorw, P.; Radpongsangkaha, P.: Spontaneous subdural hemorrhage in infants beyond the neonatal period. J. trop. Paediat. *23:* 226 (1977).

99 Iizuka, A.; Nagao, T.; Miyama, J.: Severe bleeding tendency due to prothrombin complex deficiency in young infants. Jap. J. Pediat. *23:* 226 (1977).

100 Hanawa, Y.; Maki, M.; Murata, B.; Matsuyama, E.; Yamamoto, Y.; Nagao, T.; Yamada, K.; Ikeda, I.; Terao, T.; Mikami, S.; Shiraki, K.; Komazawa, M.; Shirahata, A.; Tsuji, Y.; Motohara, K.; Tsukimoto, I.; Sawada, K.: The second nation-wide survey in Japan of vitamin K deficiency in infancy. Eur. J. Pediat. *147:* 472 (1988).

101 Lorber, J.; Lilleyman, J.S.; Peile, E.B.: Acute infantile thrombocytosis and vitamin K deficiency associated with intracranial hemorrhage. Archs Dis. Childh. *54:* 471 (1979).

102 Cooper, N.A.; Lynch, M.A.: Delayed hemorrhagic disease of the newborn with extradural hematoma. Br. Med. J. *278:* 164 (1979).

103 Minford, A.M.B.; Eden, O.B.: Haemorrhage responsive to vitamin K in a 6-week-old infant. Archs Dis. Childh. *54:* 310 (1979).

104 Nammacher, M.A.; Willemin, M.; Hartmann, J.R.; Gaston, L.W.: Vitamin K deficiency in infants beyond the neonatal period. J. Pediat. *76:* 549 (1970).

105 O'Conner, M.E.; Livingstone, D.S.; Hannah, J.; Wilkins, D.: Vitamin K deficiency and breast feeding. Am. J. Dis. Child. *137:* 601 (1983).

106 Lane, P.A.; Hathaway, W.E.; Githens, J.H.; Drugan, R.D.; Rosenberg, D.A.: Fatal intracranial hemorrhage in a normal infant secondary to vitamin K deficiency. Pediatrics *72:* 562 (1983).

107 Sutor, A.H.; Pancochar, H.; Niederhoff, H.; Pollman, H.; Hilgenberg, F.; Palm, D.; Kunzer, W.: Vitamin-K-Mangelblutungen bei vier voll gestillten Säuglingen im Alter von 4–6 Wochen. Dt. med. Wschr. *108:* 1653 (1983).

108 Widdershoven, J.; Motohara, K.; Endo, F.; Matsuda, I.; Monnens, L.: Influence of the type of feeding on the presence of PIVKA-II in infants. Helv. paediat. Acta *41:* 25 (1986).

109 Verity, C.M.; Carswell, F.; Scott, G.: Vitamin K deficiency causing infantile intracranial hemorrhage after the neonatal period. Lancet *i:* 1439 (1983).

110 McNinch, A.W.; Upton, C.; Samnels, M.; Shearer, M.J.; McCarthy, P.; Tripp, J.H.; Orme, R.L.E.: Plasma concentration after oral or intramuscular vitamin K_1 in neonates. Archs Dis. Chidlh. *60:* 814 (1985).

111 Matsuda, I.; Nishiyama, S.; Motohara, K.; Endo, F.; Ogata, T.; Futagoishi, Y.: Late neonatal vitamin K deficiency associated with subclinical liver dysfunction in breast fed infants. J. Pediat. *114:* 602 (1989).

112 National Research Council Food and Nutrition Board: Recommended Dietary Allowance, 9th ed. (National Academy of Science, Washington 1980).

113 Kries, P.V.; Kreppel, S.; Becker, A.; Tangermann, R.; Gobel, U.: Acarboxy-prothrombin activity after oral prophylactic vitamin K. Archs Dis. Childh. *62:* 938 (1987).

114 Dunn, P.M.: Vitamin K for all newborn babies. Lancet *ii:* 770 (1982).

115 Motohara, K.; Endo, F.; Matsuda, I.: Vitamin K deficiency in breast fed infants at one month of age. J. Pediat. Gastroenterol. Nutr. *5:* 931 (1986).

116 Matsuda, I.; Motohara, K.; Endo, F.; Ogata, T.: Vitamin K prevention of neonate and late neonatal bleeding. Acta paediat. jap. *31:* 436 (1989).

Ichiro Matsuda, MD, PhD, Department of Pediatrics, Kumamoto University Medical School, Kumamoto 860 (Japan)

Simopoulos AP (ed): Selected Vitamins, Minerals, and Functional Consequences of Maternal Malnutrition. World Rev Nutr Diet. Basel, Karger, 1991, vol 64, pp 109–138

Comparative Properties of Erythrocyte Calcium-Transporting Enzyme in Different Mammalian Species

Enitan A. Bababunmi[a], *Olufunso O. Olorunsogo*[a], *Clement O. Bewaji*[b]

[a] Biomembrane Research Laboratories, Department of Biochemistry, College of Medicine, University of Ibadan, Nigeria, and
[b] Biochemistry Department, University of Ilorin, Nigeria

Contents

1. Introduction

Specific physiological requirements by animals are known for at least sixteen metallic elements including the nutritionally important minerals (Ca, K, Mg, and Na), seven trace metallic elements (Mn, Fe, Co, Cu, Zn, Mo, and Se), and seven other metals which are required for animal nutrition but which have no essential functions in humans (Cd, Ni, Si, Sn, V, and As).

Calcium is the preponderant mineral element in the human body. In adult life, the human body normally contains about 1.5 kg of the metal 99% of this quantity being in the skeleton. The recommended daily dietary allowance for calcium is about 1 g in adults. During the last 22 years, Ca^{2+} has become recognized as a key messenger in the regulation of a variety of cellular processes such as muscle excitation and contraction. The study of calcium transport in mitochondria and sarcoplasmic reticulum and across plasma membranes is particularly important from the standpoint of understanding calcium homeostasis in the cell; the free cytoplasmic calcium concentration in muscle and other cells is maintained at a much lower concentration than in the extracellular fluid.

A low intracellular free Ca^{2+} concentration makes the use of the ion as an intracellular messenger energetically inexpensive. The transport of Ca^{2+} ions across biological membranes (plasma or intracellular) requires energy supplied in the form of ATP. Thus, if the resting intracellular concentration of free Ca^{2+} were large, a large amount of the cation would need to be moved into the cell to increase the concentration several fold, in order to enhance an enzyme that requires Ca^{2+} for activity. It seems clear therefore, that the normal very low free Ca^{2+} concentration implies that very few numbers of ions would be moved to regulate an enzyme. Furthermore, a low intracellular free Ca^{2+} concentration seems a necessary prerequisite for the phosphate-driven metabolism that is characteristic of higher organisms. This is more so, because if the intracellular free Ca^{2+} concentration were to be high, this cation will combine readily with the inorganic phosphate that is released from ATP hydrolysis during an ATP-driven reaction, to form a precipitate of hydroxyapatite crystals – the same stony substance found in bone.

The exact mechanism of regulating intracellular free Ca^{2+} concentration is now becoming clear. However, it has become considerably difficult to determine the mechanism of catalysis of the Ca^{2+}-translocating ATPase, the enzyme responsible for regulating intracellular Ca^{2+} concentration in erythrocytes, because of the problems of determining the amino acid sequence of the pump protein which represents only 0.1% of the total erythrocyte membrane protein.

2. Calcium and Cell Functions

The role of calcium in cell functions was first recognized as a result of the pioneering studies of Ringer [1882, 1883] who presented evidence that calcium ions (Ca^{2+}) were important in the contraction of frog heart muscle. These studies were later confirmed and extended by Locke [1894]. It took almost sixty years before the thread of research was picked up again by Heilbrunn and Wiercinski [1947] who showed that the injection of a small amount of Ca^{2+} into a muscle fibre causes it to contract.

Ca^{2+} is now known to mediate a wide variety of cellular responses and processes such as cell motility, cytoplasmic streaming, endo- and exocytosis, and more complicated processes as cell proliferation fertilization and hormone secretion (table 1). However, because of the paucity of information about calcium receptors, the mechanism of action of Ca^{2+} in most of the processes in which it has been implicated is still largely unknown.

The first substantial information on the molecular mechanisms by which Ca^{2+} signals act came from studies on the regulation of muscle contraction. In 1967, Ebashi and his co-workers demonstrated the presence of a protein that enables calcium to trigger contraction in skeletal muscle and

Table 1. Calcium-dependent cellular reactions and processes

Enzyme/process	Reference
Phosphorylase kinase (glycogenolysis)	Cohen et al. [1979]
Phospholipase A_2	Wong and Cheung [1979]
Myosin light chain kinase	Walsh et al. [1980]
Erythrocyte Ca²⁺-ATPase	Schatzmann [1966]
Adenylate cyclase	Brostom et al. [1977]
Phosphodiesterase	Cheung [1970]
Cell motility	Tash and Mann [1973]
Muscle contraction	Ebashi et al. [1967]
Exo- and endocytosis	Linden et al. [1981]
Cell division and proliferation	Welsh et al. [1978]
Fertilization	Epel et al. [1981]

Table 2. Intracellular Ca^{2+} receptor proteins

Protein	Ca^{2+}-binding sites	Reference
Calmodulin	4	Klee and Vanaman [1982]
Regulatory myosin light chain	1	Kretsinger [1980]
α-Parvalbumin	2	Haiech et al. [1979]
β-Parvalbumin	2	Coffee and Bradshaw [1973]
Skeletal muscle troponin C	4	Kretsinger [1980]
Cardiac muscle troponin C	3	Seeman and Kretsinger [1983]
Essential myosin light chain	0	Seeman and Kretsinger [1983]
Calcineurin B	4	Szebenyi et al. [1981]
Oncomodulin	2	MacManus et al. [1983]
Brain Ca^{2+}-binding protein (Caligulin)	1	Tanaka et al. [1982]

named it troponin. This protein was subsequently shown to be a complex of three polypeptides which act in concert as a Ca^{2+}-dependent trigger of muscle contraction [Ebashi and Endo, 1968]. One of the subunits, troponin C, was shown to bind Ca^{2+} with high affinity, inducing a conformational change that alters the interactions within the troponin complex. Consequently, the interactions between the troponin complex and the tropomyosin actin thin filament are also altered, permitting actin to activate myosin ATPase and produce contraction.

In recent times, investigation into the functions and mechanism of action of calcium at the molecular level has gained widespread attention among researchers in almost all disciplines of biology and medicine. The pace of research has been particularly intense during the past decade, primarily as a result of the realization that a homologous class of Ca^{2+}-binding proteins, detectable with the cationic dye Stains-All and of which calmodulin is the most ubiquitous, serve as Ca^{2+} receptors and mediate calcium functions. Most eukaryotic cells contain these Ca^{2+}-binding proteins which are structurally related to calmodulin but are tissue- and species-specific. These proteins summarized in table 2 interact reversibly with Ca^{2+} to form a protein*/Ca^{2+} complex, whose activity is regulated by the cellular flux of Ca^{2+} [Cheung, 1980]. Consequently, the responses of Ca^{2+} in a given tissue can be mediated by calmodulin but also depend on the amount and distribution of the various calmodulin target proteins and other Ca^{2+}-binding proteins.

It is now well established that calmodulin exerts its biological effects directly by a Ca^{2+}-dependent interaction with the target enzyme such as cyclic nucleotide metabolizing enzymes, plasma membrane ATPase, ciliary dyne in ATPase and cytoskeletal proteins or indirectly by activating protein kinases or phosphatases.

2.1. The Role of Calcium as an Intracellular Messenger

It is now generally accepted that Ca^{2+} is a very important, perhaps the most important second messenger in living cells. It has also been suggested that Ca^{2+} may play the role of a primary messenger, since it directly interferes with the generation of signals at the level of plasma membranes by regulating K$^+$, Na$^+$ and even Ca^{2+} currents [Carafoli and Crompton, 1978a]. In addition to its universal messenger role, Ca^{2+} is thought to act as (1) a minatory messenger because excess Ca^{2+} results in cell death or cell dysfunction; (2) a mercurial messenger in that an elevation in its intracellular concentration is short-lived and (3) a synarchic messenger as it regulates cell function in concert with another intracellular messenger, cyclic AMP (c-AMP).

In multicellular organisms, each cell must coordinate the wide range of activities within its cytoplasm in concert with those of neighbouring cells. The need for intercellular communication is met by a set of messenger molecules and cell-surface receptors that transduce the chemical messanges into a recognizable signal. The signal either activates or inhibits a biochemical reaction that is controlled by a rate-limiting step, which is usually governed by a cellular regulator – a molecule that controls one or more critical processes.

Hormones, cyclic nucleotides, and calcium are the three most important regulators or messengers in mammalian systems, and their activities are interwoven. The concept that many hormones act via a second or intracellular messenger was first put forward by Sutherland et al. [1968]. It was based on the observation that one of the early manifestations of hormonal action is a rise in the level of cAMP within the cell. Subsequent experimental evidence provided additional support for this hypothesis. It was further postulated that rather than serving as distinct and separate messengers, the functions of Ca^{2+} and cAMP were interrelated [Rasmussen, 1970; Robinson et al., 1971; Rasmussen and Goodman, 1977].

A major feature of second messenger systems is the presence of (1) a sensor which is capable of binding and releasing the messenger molecule. When bound to the messenger, the sensor molecule undergoes a conformational change and propagates information to the other macromolecules, thereby triggering a physiological response; (2) a suppressor that removes the messenger from the cytosolic compartment either by binding it after a kinetic delay or by extruding it through a membrane at the expense of ATP hydrolysis.

For the Ca^{2+} system, the sensors are members of a family of Ca^{2+}-binding proteins, the best characterized of which are skeletal muscle troponin C [Perry, 1979], and calmodulin, the ubiquitous and multifunctional Ca^{2+} receptor in non-muscle cells [Cheung, 1980; Klee and Vanaman, 1982; Babu et al., 1985]. The suppressor molecules are either the membrane-bound $(Ca^{2+} + Mg^{2+})$-ATPase, which extrudes Ca^{2+} from the cytosolic compartment, or soluble Ca^{2+}-binding proteins such as parvalbumin of fast skeletal muscle and nervous tissue [Baron et al., 1975], which remove Ca^{2+} from the sensor molecules faster than the Ca^{2+} pumps could.

2.2. The Regulation of Intracellular Calcium

Ca^{2+} ions enter cells by passive diffusion down its concentration gradient through calcium channels or molecular pores. In excitable cells, these channels open only in response to the action potential which is triggered on hormone binding at its receptor site of the cell surface. In non-excitable cells, calcium channels are not voltage-dependent and may therefore remain open permanently.

A consequence of the messenger role of a Ca^{2+} is the necessity for its precise regulation. The control of the intracellular level of Ca^{2+} is also an essential step in metabolic regulation. It is widely accepted that in mammalian cells, the steady-state concentration of Ca^{2+} is about 0.1 μM. This is four orders of magnitude lower than the free Ca^{2+} concentration in the extracellular environment, which has been estimated to be about 1.5 mM [Rasmussen and Goodman, 1977; Carafoli and Crompton, 1978a].

It is well known that a low cytosolic free Ca^{2+} concentration in erythrocytes is brought about as a result of (1) a low Ca^{2+} permeability of the erythrocyte membrane [Porzig, 1972; Ferreira and Lew, 1975]; (2) the presence of a large number of intracellular anions and high affinity membrane binding sites which are able to bind or chelate Ca^{2+} [Edmondson and Li, 1976; Chau-Wong and Seeman, 1971; Kretsinger, 1975]. Some of these ligands are simple compounds such as adenine nucleotides, inorganic phosphate and citrate. Others are proteins which are able to bind Ca^{2+} with high specificity and affinity [Kretsinger, 1975; Carafoli and Crompton,

Table 3. Calcium transporting systems in biological membranes

Transport system	Membrane types
ATPases	plasma membranes sarcoplasmic reticulum endoplasmic reticulum
Exchangers	plasma membranes (Na$^+$/Ca^{2+}) inner mitochondrial membrane (Na$^+$/Ca^{2+}; H$^+$/Ca^{2+})
Channels	plasma membranes
Electrophoretic uniporters	inner mitochondrial membranes

1978b]. However, these ligands may not be effective in the modulation of the cytosolic free Ca^{2+} concentrations in concert with physiological demands. The maintenance of intracellular Ca^{2+} at a level much lower than in the extracellular medium must depend on its extrusion through the plasma membrane, to counteract the continuous passive influx, rather than on sequestration within intracellular organelles or binding to other ligands, and (3) apart from the calcium pump, three other systems are responsible for the crucial removal of Ca^{2+} from the cytosol of mammalian cells other than the red cells. These systems are the mitochondria, the endoplasmic reticulum and the Na$^+$/Ca^{2+} exchanger in the plasma membrane.

In summary, there are four basic mechanisms by which Ca^{2+} is transported across biomembranes: Ca^{2+} channels, Ca^{2+}-pumping ATPases, exchangers, and electrophoretic uniporters (inner mitochondrial membranes) (table 3). In most eukaryotic cells, the same membrane system may transport Ca^{2+} by more than one of these mechanisms [Carafoli, 1987]. Ca^{2+}-pumping ATPases have so far been described in plasma membranes and in sarcoplasmic reticulum. Similarly, Na$^+$/Ca^{2+} exchangers have been reported in plasma membranes and the inner mitochondrial membrane.

2.3. Calcium Transport across the Plasma Membrane

Three Ca^{2+}-transporting systems which mediate the movement of Ca^{2+} across various plasma membranes have so far been described. These are: the Ca^{2+} channel, the Ca^{2+}-pumping ATPase, and the Na$^+$/Ca^{2+} exchanger.

2.3.1. The Ca^{2+} Channel

The activities of the Ca^{2+} channel, which mediate the influx of Ca^{2+} into cells, were first studied in excitable plasma membrane. Ca^{2+} action potentials, implying a Ca^{2+} component in plasma membrane conductance were recorded for the crayfish muscle fibre membrane by Fatt and Ginsborg [1958]. The Ca^{2+} component was subsequently attributed to the existence of a specific Ca^{2+} channel. Ca^{2+} channels are usually investigated by recording electrical currents in whole tissue or cell preparations. Quite recently, studies on the Ca^{2+} channel have been extended to non-excitable cells where currents cannot be measured [Varecka and Carafoli, 1982]. In these cells, other criteria for the definition of the Ca^{2+} channels have been employed such as saturation kinetics of the Ca^{2+} transport process, competitive inhibition by ions like Ca^{2+}, and inhibition by the Ca^{2+}-entry blockers.

The classical Ca^{2+}-channel inhibitors are the Ca^{2+}-entry blockers which constitute a large class of agents that can be divided into two subcategories: tertiary amines (verapamil, diltiazem), and dihydropyridines (nifedipine, nitrendipine). The former are thought to act on the channel from the inside, and must therefore penetrate into the cell to exert their effects. Thus, their inhibition is voltage-dependent. The latter are thought to act on the channel from the outside and are thus voltage-insensitive [Reuter et al., 1982; Reuter, 1984; Carafoli, 1987].

Recently, considerable advances in the characterization of the Ca^{2+} channel have been made possible by the development of the voltage-patch technique [Hamill et al., 1981].

2.3.2. The Ca^{2+}-Pumping ATPase

The existence of a Ca^{2+}-stimulated, Mg^{2+}-dependent ATPase in plasma membranes was first reported by Dunham and Glynn [1961] who noted that the simultaneous presence of Mg^{2+} and Ca^{2+} increases the total ATP hydrolysis by isolated membranes from human erythrocytes, and at the same time inhibits the Na^+, K^+-ATPase. This finding was confirmed by Hoffman [1962] who concluded that the site of Ca^{2+} inhibition of the Na^+, K^+-ATPase is on the inner side of the membrane.

The presence of an active Ca^{2+} pump in human red blood cells was first observed by Schatzmann [1966] who demonstrated that the Ca^{2+}-pump mechanism was closely coupled to a Mg^{2+}-dependent, Ca^{2+}-stimulated adenosine triphosphatase $(Ca^{2+} + Mg^{2+})$-ATPase [Schatzmann and Vincenzi, 1969].

The Ca^{2+}-ATPase has been extensively studied in erythrocytes by several investigators mainly because the plasma membranes of these cells can be obtained in pure form uncontaminated by intracellular organelles. It is now generally accepted that it is the enzyme responsible for maintaining the concentration of Ca^{2+} inside the cell at levels much lower than in the surrounding medium [Schatzmann, 1982]. Earlier attempts to study the properties of the enzyme in unfractionated erythrocyte ghosts were complicated due to the presence of a Mg^{2+}-ATPase in the membrane. Purification of the Ca^{2+}-ATPase also proved difficult as a result of the lability of the enzyme, its low concentration in the membrane, and the fact that its molecular weight and solubilization properties are similar to those of band 3 – one of the most abundant components of the erythrocyte membrane [Ronner et al., 1977]. However, the enzyme was subsequently purified by applying affinity chromatographic techniques which involved the coupling of calmodulin to a Sepharose 4B matrix [Niggli et al., 1979b; Gietzen et al., 1980].

Results of studies from several laboratories have shown that the values obtained for the activity and Ca^{2+}-sensitivity of the Ca^{2+}-pumping ATPase differ depending on the method of membrane preparation, the nature of the cofactors in the reaction medium and the presence or absence of regulators [Roufogalis, 1979]. Earlier studies demonstrated that the $(Ca^{2+} + Mg^{2+})$-ATPase activity in red cell membranes obtained by hypotonic lysis was stimulated by Ca^{2+} in a biphasic manner [Schatzmann and Rossi, 1971; Scharff, 1972; Horton et al., 1970]. The activity stimulated by concentrations of calcium below 10 μM was called high Ca^{2+}-affinity component while the activity stimulated by Ca^{2+} concentrations in the range of 10–300 μM was termed the low Ca^{2+}-affinity component [Schatzmann and Rossi, 1971]. Two years later, Bond and Clough [1973] demonstrated an enhancement of the $(Ca^{2+} + Mg^{2+})$-ATPase activity by the addition of a heat stable factor in the haemolysate obtained from the lysis of erythrocytes. Quist and Roufogalis [1975] showed that although the contribution of the high affinity component is diminished on incubation of ghost membranes with 0.1 M EDTA at 37 °C, this affinity is however, restored on readdition of the material extracted by EDTA to the ghost membranes. The heat-stable cytosolic activating factor was later shown to be calmodulin. Calmodulin has also been reported to stimulate the ATP-dependent Ca^{2+} uptake into inside-out vesicles prepared from human erythrocytes [Larsen and Vincenzi, 1979].

The specific activities reported for the purified Ca^{2+}-ATPases preparations range from 9.0 to 186 μmol/mg protein/h, compared with the usual values of 0.3–3.0 μmol/mg protein/h for whole membranes [Gietzen et al., 1980; Niggli et al., 1981a]. The purified enzyme preparation, when reconstituted into artificial liposomes, is activated by calmodulin about 7-fold, in analogy with the findings on the membrane-bound enzyme. Calmodulin shifts the purified and membrane-bound ATPase from a low Ca^{2+}-affinity state (K_m for Ca^{2+} about 10–14 μM) to a high Ca^{2+}-affinity state (K_m for Ca^{2+} about 1 μM) [Niggli et al., 1981a].

Various delipidation experiments have been used to show that the Ca^{2+}-ATPase requires phospholipids for activity [Roelofsen and Schatzmann, 1977; Ronner et al., 1977; Peterson et al., 1978]. Niggli et al. [1981b] have also shown that calmodulin is not unique as an activator of the enzyme. These workers showed that the transition from a low to the high Ca^{2+}-affinity state can be induced, in the absence of calmodulin, by a variety of acidic phospholipids (phosphatidylserine, phosphatidylinositol and phosphatidic acid). These phospholipids increase the V_{max} of the enzyme and its affinity for Ca^{2+} to the same extent as calmodulin. The purified and reconstituted enzyme, in the absence of calmodulin, can also be stimulated by controlled proteolysis with trypsin or chymotrypsin [Taverna and Hanahan, 1980; Niggli et al., 1981b; Al-Jobore and Roufogalis, 1981]. The proteolyzed enzyme has the same V_{max} and high affinity for Ca^{2+} as the untreated enzyme in the presence of calmodulin. It has been suggested that the essential machinery of the red cell Ca^{2+} pump is contained in a single polypeptide with a molecular weight of about 140,000 daltons but this is normally turned off until it encounteres the right Ca^{2+}/*calmodulin complex. Interaction with anionic amphiphilic molecules, whether fatty acids or phospholipids, may somehow mimic the effect of calmodulin, and proteolysis may remove the control site, leaving the enzyme fully active [Stieger and Schatzmann, 1981; Niggli et al., 1981b; Michell, 1982].

Quite recently, it has been reported that polyphosphoinositides are also powerful activators of the Ca^{2+}-ATPase of erythrocyte membranes [Penniston, 1982, 1983; Choquette et al., 1984]. These phospholipids are thought to be of importance in the regulation of the enzyme. It has been suggested that phospholipids, fatty acids and the Ca^{2+}/*calmodulin complex may activate the enzyme by introducing a relatively hydrophobic environment in the neighbourhood of the active site of the enzyme. This would produce a conformational change in the protein molecule in such a way as to make the active site more accessible. The proteolytic treatment would produce the same effect by removing a hypothetical inhibitory peptide in the neighbourhood of the active site [Carafoli and Zuruni, 1982].

2.3.3. The Na⁺/Ca²⁺ Exchanger

The observation in a number of excitable tissues that the efflux of Ca^{2+} against a concentration gradient was dependent on the presence of Na^+ in the external medium, led to the discovery of an exchange diffusion that couples the fluxes of Na^+ and Ca^{2+} across the plasma membrane [Reuter and Seitz, 1968; Blaustein and Hodgkin, 1969]. Subsequent experiments have shown that this carrier, now known as the Na^+/Ca^{2+} exchanger, may also mediate the influx of Ca^{2+} into cells [Baker et al., 1969] and that it exists in nonexcitable plasma membranes as well. The erythrocyte may be the only eukaryotic membrane, among those studied, where the exchanger has not been demonstrated. However, it has been reported that dog erythrocytes possess a Na^{2+}-coupled Ca^{2+} transport system, but its properties are different from those of the exchanger [Parker, 1978].

The work of several investigators on heart sarcolemmal vesicles, reviewed by Blaustein and Nelson [1982], have shown that the exchanger operates electrogenically with a stoichiometry of $2Na^+$ for $1\ Ca^{2+}$. These experiments have also permitted the estimation of some important kinetic parameters of the exchange process. Reeves and Sutko [1979] have found that K_m for Ca^{2+} is about 20 μM and for Na^+ about 15 mM. They have also measured a V_{max} of approximately 0.4 nmol of Ca^{2+} transported per milligram of sarcolemmal protein per second. More recent experiments in which the constraints imposed on the reaction by its electrogenicity were eliminated [Caroni et al., 1980] have established that the affinity of the exchanger for Ca^{2+} is much higher (K_m about 2 μM) and its maximal velocity of Ca^{2+} pumping is considerably faster (up to 30 μmol/mg sarcolemmal protein per second). Thus, the exchanger appears to be a high capacity Ca^{2+} transporting system which interacts with Ca^{2+} with a lower affinity than the Ca^{2+}-pumping ATPase.

3. Calcium Transport across Intracellular Membranes

3.1. Sarcoplasmic Reticulum

Calcium plays an important role in the regulation of muscle contraction. Skeletal muscle contains an intricate network of membranes called the sarcoplasmic reticulum which regulates the Ca^{2+} concentrations in the medium surrounding the contractile fibres of muscle.

The active transport of Ca^{2+} across the membranes of sarcoplasmic reticulum has received much attention for the past several years from biochemical and biophysical points of view. The protein composition of this membrane system has been extensively investigated and four major proteins have been identified [Meissner et al., 1973]. All were considered to be related, to some extent, to the Ca^{2+} transport activity. The ATPase protein with a molecular weight of about 100,000 daltons accounts for up to 90% of the total protein. Other proteins are calsequestrin, high-affinity calcium-binding protein, and a proteolipid which is now known as phospholamban [Tada et al., 1975]. There are also some minor proteins whose properties are not well defined [MacLennan et al., 1973].

Reports on rabbit skeletal muscle sarcoplasmic reticulum indicate that the ATPase accounts for 60–80% of the total membrane protein, calsequestrin accounts for 5–19% and the high-affinity calcium-binding protein for 5–12% [Meissner et al., 1973]. However, Meissner and Fleischer [1971] reported that sarcoplasmic reticulum of high and low density, separated on a sucrose gradient, had different compositions. They reported that the light vesicles were composed largely of ATPase (about 90%) whereas the heavy vesicles contained ATPase (55–60%), calsequestrin (20–25%), and the high-affinity calcium-binding protein (5–7%).

The ATPase has been purified by the use of mild detergents [MacLennan, 1970]. Its molecular weight is in the range 100,000–120,000 daltons. Indirect evidence suggests that the enzyme forms oligomers within the membrane, presumably tri- or tetramers [Le Maire et al., 1976]. It has been observed that a phosphorylated protein intermediate is formed during the process of translocation of Ca^{2+} into the vesicles [Tada et al., 1978].

It has also been shown that the substrate of the enzyme is MgATP [Vienna, 1975]. The hydrolysis of ATP is initiated by the transfer of the γ-phosphate of ATP to an aspartyl residue of the enzyme. This step depends on the binding of Ca^{2+} to a high affinity site on the enzyme located at the outer surface of the membrane. Ca^{2+} is translocated through the membrane and released in the vesicular lumen prior to the hydrolysis of the phosphoenzyme. Hasselbach and Makinose [1962] have shown that the number of Ca^{2+} ions translocated per ATP hydrolyzed is 2.

Tada et al. [1978] have proposed a reaction mechanism to illustrate the sequence of reactions leading to ATP hydrolysis and Ca^{2+} transport (fig. 1). In the E form, the Ca^{2+}-binding site faces the outer surface of the vesicle and has an apparent K_m for Ca^{2+} in the range 0.2–2.0 μM at pH 7.0 (high affinity). In the *E form, the Ca^{2+}-binding site faces the inner surface of the vesicle and has an apparent K_m for Ca^{2+} in the range 1–3 mM at pH 7.0 (low affinity). The E form is phosphorylated by ATP but not by inor-

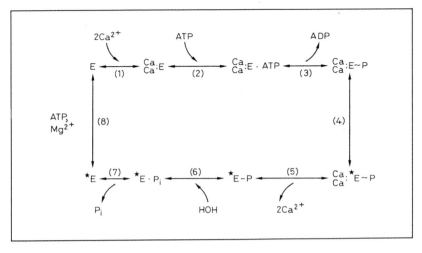

Fig. 1. The enzyme cycle of the sarcoplasmic reticulum Ca^{2+}-ATPase. The reaction sequence is explained in the text. E and *E represent different conformational states of the enzyme. From Tada et al. [1978].

ganic phosphate, while the *E form is phosphorylated by inorganic phosphate and not by ATP.

Attempts have been made to explain the mechanism of active cation transport across the sarcoplasmic reticulum by assuming that the membrane ATPase serves as an energy transducer as well as a translocator of cations. Kinetic analysis of the ATPase suggests that the enzyme forms an enzyme-substrate-calcium complex at the outer surface of the membrane, followed by the formation of a phosphorylated intermediate that is coincident with the translocation of Ca^{2+} from outside to inside of the membrane. The phosphorylation of the enzyme is associated with a dramatic change in the affinity for Ca^{2+} and Mg^{2+}, so that Mg^{2+}, having increased affinity for the phosphorylated intermediate, accelerates its decomposition. The structural basis for the profound change in the affinities of the enzyme for cations is not yet clearly understood [Tada et al., 1978].

3.2. Mitochondria

The accumulation of Ca^{2+} by respiring mitochondria was first reported by Vasington and Murphy [1961, 1962] and by DeLuca and Engstrom [1961].

Mitochondria from almost all tissues that have been studied are known to accumulate calcium from the surrounding medium. These tissues include liver, heart, skeletal and smooth muscles, kidney, spleen, brain and thyroid [Lehninger et al., 1967; Carafoli and Lehninger, 1971; Wikstrom et al., 1975].

The energy-dependent accumulation of Ca^{2+} by mitochondria is driven by a membrane potential created by respiration of ATP hydrolysis. In the former case, the process is inhibited by respiratory chain inhibitors, and in the latter case by oligomycin. In both cases, it is inhibited by uncouplers [Lehninger et al., 1967; Carafoli and Crompton, 1976; Bygrave, 1977].

The detailed mechanisms of this process have been unravelled as a result of studies in which mitochondria were shown to accumulate very limited amounts of Ca^{2+}, in the absence of inorganic phosphate. It is now known that Ca^{2+} uptake is an electrophoretic process driven by the transmembrane electrical potential maintained by respiration [Rottenberg and Scarpa, 1974] and inhibited by ruthenium red. The uptake process has a rather low affinity for Ca^{2+} (K_m = 2–13 μM). Its maximal transport rate, which is optimal in the presence of inorganic phosphate, is of the order of 10 nmol/mg protein at 25 °C. It is also inhibited by physiological concentrations of Mg^{2+}, implying that under physiological conditions, where Mg^{2+} is present, and the concentration of Ca^{2+} is well below the K_m value, the rate of Ca^{2+} uptake is only a fraction of the maximal velocity.

A very interesting development in this field was the discovery that the uptake and release of Ca^{2+} by mitochondria take place via different routes and mechanisms [Crompton et al., 1976]. This conclusion was based on theoretical considerations on the equilibrium of the Ca^{2+} transport process in the presence of normal transmembrane potential maintained by mitochondrial respiration, and on the finding that calcium can be released from mitochondria in the presence of ruthenium red, which blocks the uptake pathway.

Lehninger et al. [1978] have also shown that calcium-preloaded mitochondria from various tissues released calcium in response to the addition of aceto-acetate or oxalo-acetate. This release was reversed by the addition of β-hydroxybutyrate. These results demonstrate that the release of calcium from mitochondria is regulated by the oxidation-reduction state of pyridine nucleotides, thus regulating its concentration in the external medium, the cytosol. The mitochondria used were maintained in an energised state throughout the experiment by the addition of succinate plus rotenone or other pyridine nucleotide-independent substrates. It was also shown that the phenomenon was ruthenium red-insensitive, suggesting that the efflux of calcium, activated by the shift in the redox state of pyridine nucleotides, followed a route different from the uptake pathway.

4. An Integrated Picture of Calcium Homeostasis

The role of the various calcium transporting systems in intracellular calcium homeostasis seems rather complex. The regulation of intracellular calcium is the direct result of the cooperation between these systems. In addition, it has been suggested that various non-membranous ligands may also contribute to the overall process of calcium homeostasis [Carafoli and Crompton, 1978b].

In the heart, the kinetic parameters of the four membranous Ca^{2+} transporting systems have been well defined, since they are all present in this organ [Carafoli, 1987]. It is generally accepted that under physiological conditions, only minor fractions of the intracellular calcium ions are transported by the plasma membrane and mitochondria; the most active organelle is the sarcoplasmic reticulum [Bygrave, 1977; Carafoli, 1987; Lehninger et al., 1967; Vianna, 1975]. Mitochondria are regarded as long-term calcium buffers which can handle very large amounts of calcium in emergencies which usually arise when cells are stimulated.

5. Red Cell Membrane Calcium Pump in Different Nutritionally Classified Animal Species

Schatzmann [1982] presented an excellent review on certain properties of the calcium pump of the erythrocytes of some classes of animals.

5.1. Avian Red Cells

An ATP-dependent Ca^{2+}-pumping ATPase with a very high V_{max} has been detected in vesicles made from pigeon red cells [Ting et al., 1979]. The ATPase also has a high affinity for Ca^{2+} ($K_m(Ca^{2+}) = 0.18 \, \mu M$) and a dependence of rate on Ca^{2+} concentration which could be described by a straight-line relation between reciprocal of rate and $(Ca^{2+})^2$. This latter finding is a strong evidence for the existence of the two Ca^{2+}-binding sites on the Ca^{2+} pump.

5.2. Red Cells of Carnivores

Brown [1979] and Parker [1979] have presented good evidence for the existence of a Mg^{2+}- and ATP-dependent Ca^{2+} extrusion mechanism similar to what has been described for human red cells. It has been shown that dog red cells have a high intracellular Na^+ and low K^+ concentration which has been attributed to the absence of a Na^+,K^+ pump [Parker et al., 1975; Parker, 1977, 1978].

These workers have also shown that the Ca^{2+} gradient which exists across the dog red cell membrane moves some Na^+ out of the cells. The Ca^{2+} pump thus keeps the cells from reaching electrochemical equilibrium with respect to Na^+, which would inevitably lead to swelling and eventual osmotic haemolysis. Thus, in species which lack the Na^+,K^+ pump (the dog and probably the cat), the Ca^{2+} pump maintains the normal volume of their red cells.

5.3. Ruminant Red Cells

Ruminants (sheep, goats and cattle) have red cells which display three special features with respect to cation transport:

(1) While it is generally accepted that K^+ is the principal intracellular cation of mammalian erythrocytes, it has been established that the pattern of distribution of these monovalent cations is reversed in erythrocytes of ruminants [Kerr, 1937; Evans, 1954; Evans and Phillipson, 1957; Tosteson, 1963]. The difference is explained by the observation that low K^+ cells have fewer Na^+,K^+ pump sites per cell and, more importantly, that their pump has a higher affinity for K^+ on the internal surfaces of the membrane, leading to a competition between Na^+ and K^+ for the binding site at considerably lower internal K^+ levels than in human red cells.

(2) Ruminant red cells lack the Ca^{2+}-sensitive K^+ channel after the foetal period [Jenkins and Lew, 1973; Brown et al., 1978].

(3) The red cell Ca^{2+},Mg^{2+}-ATPase activity in adult ruminants is considerably lower than in human red cell [Schatzmann, 1974]. It is of interest to note that foetal cells differ from adult cells in these species in that they have high intracellular K^+ concentration and a Ca^{2+}-sensitive K^+ channel which is lost in the adult [Brown et al., 1978]. They also have a high Ca^{2+},Mg^{2+}-ATPase activity [Schatzmann and Scheidegger, 1975].

6. Bovine, Porcine, Ovine and Caprine Erythrocytes

The relative amounts of erythrocyte membrane polypeptides in various mammalian species have been determined (table 4) [Bewaji et al., 1985]. The results demonstrate that there are significant differences in the erythrocyte membrane polypeptides especially in the spectrin region (bands 1 and 2). It has been shown that porcine erythrocytes have relatively greater amounts of band 2.3 than human, bovine and ovine erythrocytes [Bewaji et al., 1985], while this protein is completely absent in caprine erythrocyte membrane. The same study also suggests that band 3 protein in human erythrocytes is not as homogenous as in other species. In addition, porcine erythrocytes countain some polypeptides which are absent in other species. It is not clear if these bands represent a split of band 7 during membrane isolation rather than an occurrence in the membrane of 2 distinct bands.

Studies on the basal Ca^{2+}-translocating ATPase of the erythrocytes from human, bovine, porcine, ovine and caprine (table 5) revealed that of the five mammalian species investigated, specific activity of the porcine erythrocyte is several orders of magnitude highest ($p < 0.001$). The reason for this high value remains unknown, although we have suggested that the lipid environment of the enzyme in the native membranes of the porcine erythrocytes might be significantly different from that of other species. Buckley and Hawthorne [1972] provided the first direct evidence that the polyphosphoinositides (Ptdlns) which are now known to be localised in the plasma membranes of a wide variety of mammalian tissues, including erythrocytes, may be involved in the regulation of intracellular calcium levels. These authors demonstrated that an increase in the membrane content of these phospholipids is correlated with Ca^{2+}-pumping ATPase activity in pig erythrocytes. Because of the strong ionic properties and rapid turnover of these lipids, it has been suggested that they may also participate in the active transport of cations. Erythrocyte membranes have been shown to contain Ptdlns and Ptdlns 4-P kinases [Peterson and Kirshner, 1970] as well as a Ca^{2+}-activated phosphodiesterase that acts on phosphoinositides [Allan and Michell, 1978].

Recent work has shown that the hydrolysis of Ptdlns 4,5-P_2 is the initial event in the Ptdlns turnover and that this lipid is a potent activator of the Ca^{2+}-pumping ATPase of erythrocyte membranes and may be important in the regulation of this enzyme [Allan and Michel, 1978; Downes and Michell, 1982; Penniston, 1982]. Scheetz et al. [1982] have

Table 4. Relative amounts of erythrocyte membrane polypeptides in various mammalian species

Band	Percent of total membrane protein				
	human	bovine	porcine	ovine	caprine
1 and 2	11.99 (1.45)	26.89 (4.34)	23.63 (6.55)	16.60 (0.39)	19.40 (2.54)
2.1	2.60 (0.85)	11.26 (0.80)	5.61 (0.55)	4.49 (0.81)	12.59 (0.11)
2.2	negl.	6.59 (1.07)	negl.	6.47 (1.02)	negl.
2.3	negl.	negl.	7.86 (0.81)	negl.	–
3	25.73 (0.70)	26.00 (3.85)	26.99 (2.69)	22.40 (2.49)	29.21 (1.97)
4.1	12.27 (0.71)	2.84 (1.02)	5.50 (0.98)	7.17 (1.88)	3.21 (0.11)
4.2	12.17 (0.87)	9.36 (1.62)	8.04 (1.37)	9.43 (0.81)	11.06 (0.99)
4.5	11.14 (1.42)	8.08 (0.42)	3.00 (0.42)	4.75 (0.72)	3.69 (0.21)
5	19.20 (1.37)	15.78 (1.45)	5.35 (0.38)	13.12 (1.42)	9.69 (1.36)
6	5.19 (1.48)	5.69 (0.62)	7.25 (0.02)	5.04 (0.21)	–
7	8.98 (0.81)	–	8.01 (1.66)	5.59 (0.34)	7.60 (0.54)
8	–	–	3.88 (0.14)	5.25 (0.78)	negl.

The results represent the means of 4 determinations with SD in parentheses.
– = Not detected; negl. = negligible. From Bewaji et al. [1985].

Table 5. Effect of calmodulin on the (Ca^{2+} + Mg^{2+})-ATPase activity of erythrocyte membranes

Species	Mg^{2+}-ATPase[a]	(Ca^{2+} + Mg^{2+})-ATPase[a]	
		– calmodulin	+ calmodulin
Human	0.32 ± 0.01	0.36 ± 0.02	0.77 ± 0.02
Bovine	0.39 ± 0.01	0.40 ± 0.05	0.73 ± 0.03
Porcine	0.85 ± 0.05	2.86 ± 0.06	4.15 ± 0.35
Ovine	0.56 ± 0.02	0.73 ± 0.02	1.09 ± 0.17
Caprine	0.38 ± 0.01	0.48 ± 0.03	0.85 ± 0.04

[a] μmol inorganic phosphate liberated per milligram of membrane protein in 1 h at 37 °C. Values represent means ± SD of 4 independent experiments. From Bewaji et al. [1985].

also put forward the interesting suggestion that the fluidity of the lipid bilayer of the human erythrocyte membrane can be influenced by Ptdlns 4,5-P$_2$. These workers showed that the addition of Ptdlns 4,5-P$_2$ to membranes appeared to increase the mobility of membrane glycoproteins. We have reported that (table 6) like in human erythrocyte, phosphatidylinositol enhances the Ca^{2+}-pumping ATPase activity in bovine, porcine, ovine and caprine erythrocytes, suggesting that the mode of operation of the pump is similar in these species [Bewaji et al., 1985]. The variations in the activity of the (Ca^{2+} + Mg^{2+})-ATPase in erythrocytes from different species reported by these authors could be due to differences in the interaction between the constituent proteins and lipids. We were also able to show some quantitative differences in the membrane polypeptides which could affect erythrocyte function. For example, alteration in the amount, or complete absence of proteins which maintain erythrocyte membrane cytoskeleton could lead to inability to maintain a constant shape, ionic composition and the plasticity and elasticity which enables the erythrocyte to survive the buffering or circulation for (in the case of human erythrocytes) about 120 days. It should be noted that the erythrocytes of mice which congenitally lack spectrin – the major structural membrane protein – have a life span of 20–30 days [Eaton, 1981].

The (Ca^{2+} + Mg^{2+})-ATPase is an integral membrane protein within a molecular weight and solubilization properties similar to those of band 3 –

Table 6. Effect of phosphatidylinositol on the $(Ca^{2+} + Mg^{2+})$- ATPase activity of mammalian erythrocyte membranes

Additions	ATPase activity, μmol phospate/mg protein/h				
	human	bovine	porcine	ovine	caprine
None	0.32 (0.01)	0.37 (0.02)	1.05 (0.06)	0.59 (0.02)	0.45 (0.02)
EGTA	0.32 (0.01)	0.39 (0.01)	0.85 (0.05)	0.56 (0.02)	0.38 (0.01)
Ca^{2+}	0.36 (0.03)	0.40 (0.05)	2.86 (0.06)	0.73 (0.02)	0.48 (0.03)
Ptdlns	0.72 (0.02)	0.65 (0.07)	1.91 (0.04)	0.97 (0.02)	0.72 (0.02)
Ptdlns + EGTA	0.67 (0.04)	0.65 (0.05)	1.54 (0.07)	0.88 (0.01)	0.62 (0.01)
Ptdlns + Ca^{2+}	0.79 (0.07)	0.73 (0.05)	3.65 (0.27)	0.97 (0.04)	0.77 (0.02)

ATPase activity was assayed by measuring the inorganic phosphate released from ATP in 30 min at 37 °C as described essentially by Ronner et al. [1977]. When present, the final concentration of each additive was EGTA 500 μM, Ptdlns 40 μg/ml, Ca^{2+} 200 μM. Each value represents the mean and SD (in parentheses) of 4 separate determinations. From Bewaji et al. [1985].

Table 7. Kinetic parameters of $(Ca^{2+} + Mg^{2+})$-ATPase in erythrocyte ghosts from various mammalian species

Species	V_{max} μmol/mg protein/h	K_m(ATP), nM	$K_1(VO_4^{3-})$, μM
Human	0.38 ± 0.06	0.25 ± 0.12	6.58 ± 0.42
Bovine	0.36 ± 0.06	0.45 ± 0.18	11.84 ± 0.53
Porcine	6.25 ± 0.53	0.43 ± 0.11	2.08 ± 0.11
Ovine	1.05 ± 0.28	0.36 ± 0.12	5.63 ± 0.21
Caprine	0.42 ± 0.11	0.31 ± 0.09	8.61 ± 0.38

From Bewaji et al. [1985].

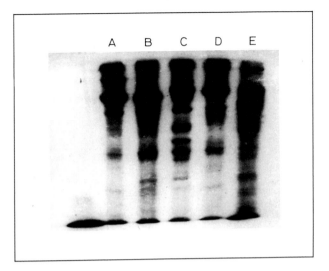

Fig. 2. SDS polyacrylamide gel electrophoresis of ghost membranes from (A) caprine, (B) ovine, (C) porcine, (D) bovine, and (E) human erythrocytes. 40 μg of ghost protein was applied to 12 % polyacrylamide slab gel and stained with Coomassie Brilliant Blue R-250.

the most abundant protein in the erythrocyte membrane [Niggli et al., 1979a]. The SDS-PAGE of porcine erythrocyte membrane proteins shown in figure 2 indicate that the proteins present in the band 2.3 region are relatively greater than in human, bovine and ovine erythrocytes. It is possible that the (Ca^{2+} + Mg^{2+})-ATPase could contribute substantially to the high amount of protein in this region. Recent work in our laboratory on the purification of the ATPase from porcine erythrocytes is suggestive in this direction. We have consistently found that the (Ca^{2+} + Mg^{2+})-ATPase in this species could constitute as high as 1 % of the total membrane proteins compared with the 0.1 % usually reported for human erythrocytes. This provides another possibility for the higher specific activity of this enzyme in porcine erythrocytes than in the other species studied.

Results of studies from our laboratory indicate that the calcium-pumping protein of the erythrocytes from various species has identical kinetic parameters (summarised in table 7) except for some minor differences. For instance, the affinity for ATP is not significantly different in all the species except in human erythrocytes which seem to have the highest

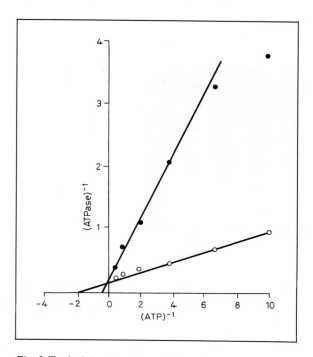

Fig. 3. Typical double reciprocal plot of (porcine) erythrocyte $(Ca^{2+} + Mg^{2+})$-ATPase in the presence (●) or absence (○) of vanadate (VO_4^{3-}). The reaction mixture contained 10 μM vanadate and ATP concentration was varied from 100 μM to 2 mM. From Bewaji et al. [1985].

affinity for ATP. V_{max} values are several orders of magnitude higher in ovine and porcine erythrocytes. The physiological implications of this is not clear. Further studies have shown that the enzyme in these species of erythrocytes is highly sensitive to vanadate. The constants of the enzyme in the various species were estimated using the Lineweaver-Burk and Dixon plots (fig. 3 and 4).

The regulation of the human erythrocyte Ca^{2+}-pumping ATPase by calmodulin is well established. Some workers have shown that the membrane-bound or purified enzyme is highly sensitive to calmodulin [Niggli et al., 1979a]. It has been shown that the membrane bound Ca^{2+}-translocating ATPase of the erythrocytes of bovine, porcine, ovine and caprine are similarly sensitive to calmodulin [Bewaji et al., 1985]. More recent studies on the purified Ca^{2+}-ATPase from pig erythrocytes indicate that the amount

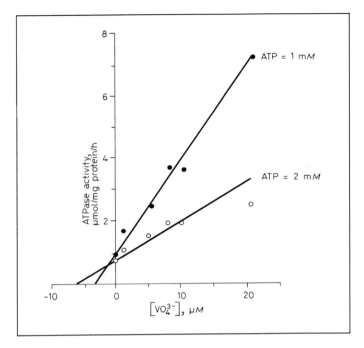

Fig. 4. Dixon plot of the inhibition of erythrocyte ghost membrane (Ca^{2+} + Mg^{2+})-ATPase activity by vanadate, using porcine erythrocyte as a model. From Bewaji et al. [1985].

of the enzyme is at least seven times greater than that isolated from human erythrocytes [Bewaji and Bababunmi, 1987]. The results also revealed that the properties of the enzyme from human and pig erythrocytes are similar in many respects. Although the abundance of the Ca^{2+}-ATPase in pig erythrocyte membranes raises the hope of obtaining enough of the enzyme, which is very similar on SDS-PAGE electrophoretic pattern [Bewaji and Bababunmi, 1987] to the human calcium-pumping protein, it is not yet clear why pig erythrocytes have such a high content of this protein. In view of the observation that the Ca^{2+}-pump in human erythrocytes appears to be functioning well below its capacity [Jarret and Kyte, 1979; Roufogalis, 1979; Penniston, 1983], it is very desirable to determine the reasons for the apparent increase in the amount of the pump protein in pig erythrocytes.

7. Conclusion and Future Direction

The Ca^{2+}-transporting ATPases are cell membrane calcium-dependent enzymes of major and universal interest. We believe in the fact that characterization of these proteins will assist us in the attempt to understand the problems of pathobiology that are very important to Africa, such as malnutrition, parasitic diseases, genetic disorders and cardiovascular diseases.

Calcium-pumping ATPases of plasma membranes are present in a wide variety of cell types such as those of the lymphatic and circulatory system, organs, tissues (excitable and non-excitable and plants) [Penniston, 1983]. Indeed, during the past decade, almost all eukaryotic cell surfaces have been shown to possess the calcium pump whose major component requires Ca^{2+} and Mg^{2+}. However, several species of the parasite Trypanosoma seem to have on their cell surfaces only the Mg^{2+} pump [Bababunmi, 1987] which is a distinct difference from the property of the ghost membrane of erythrocytes of the mammalian hosts (of different nutritional class) to trypanosomes. It is desirable to further examine the possibility of the universality of this property amongst the family of trypanosomes which are the causative agents of sleeping sickness disease of Africa and of Chagas' disease of South America. The calcium pump protein of hypertensive Nigerian Africans is stimulated by calmodulin to a lesser extent than that of the normotensive humans [Olorunsogo et al., 1985]. Therefore detailed characterization of the pump protein in certain pathological states such as hypertension, sickle cell anaemia and malnutrition should enhance our knowledge of the mechanism and molecular basis of the genesis and expression of these diseases.

References

Al-Jobore, A.; Roufogalis, B.D.: Phospholipid and calmodulin activation of solubilized calcium-transport ATPase from human erythrocytes: regulation by magnesium. Can. J. Biochem. *59:* 880–888 (1981).

Allan, D.; Michell, R.H.: A calcium-activated polyphosphoinositide phosphodiesterase in the plasma membrane of human erythrocytes. Biochim. biophys. Acta *508:* 277–286 (1978).

Bababunmi, E.A.: Plasma membrane-bound Mg^{2+}-ATPase of the bloodstream and procyclic forms of *Trypanosoma brucei*. Microbios Lett. *36:* 71–76 (1987).

Babu, Y.S.; Sack, J.S.; Greenhough, T.J.; Bugg, C.E.; Means, A.R.; Cook, W.J.: Three dimensional structure of calmodulin. Nature, Lond. *315:* 37–40 (1985).

Baker, P.F.; Blaustein, M.P.; Hodgkin, A.L.; Steinhardt, R.A.: The influence of calcium on sodium flux in squid axons. J. Physiol., Lond. *200:* 431–458 (1969).

Baron, G.; DeMaille, J.; Dutruge, E.: The distribution of parvalbumins in muscle and in other tissues. FEBS Lett. *5:* 156–160 (1975).

Bewaji, C.O.; Bababunmi, E.A.: Abundance of the Ca^{2+}-pumping ATPase in pig erythrocyte membrane. Biochem. J. *248:* 297–299 (1987).

Bewaji, C.O.; Olorunsogo, O.O.; Bababunmi, E.A.: Comparison of the membrane-bound (Ca^{2+} + Mg^{2+})-ATPase in erythrocyte ghosts from some mammalian species. Compar. Biochem. Physiol. *82B:* 117–122 (1985).

Blaustein, M.P.; Hodgkin, A.: The effect of cyanide on the efflux of calcium from squid axons. J. Physiol., Lond. *200:* 497–527 (1969).

Blaustein, M.P.; Nelson, M.: Na$^+$-Ca^{2+} exchange: its role in the regulation of cell calcium; in Carafoli, Membrane transport of calcium, pp. 217–236 (Academic Press, New York 1982).

Bond, G.H.; Clough, D.: A soluble protein activator of (Mg^{2+} + Ca^{2+})-dependent ATPase in human red cell membranes. Biochim. biophys. Acta *323:* 592–599 (1973).

Brostom, C.O.; Brostom, M.A.; Wolff, D.D.: Calcium-dependent adenylate cyclase from rat cerebral cortex. J. biol. Chem. *252:* 5677–5685 (1977).

Brown, A.M.: Evidence for a magnesium- and ATP-dependent calcium extrusion pump in dog erythrocytes. Biochim. biophys. Acta *554:* 195–203 (1979).

Brown, A.M.; Ellory, J.C.; Young, J.D.; Lew, V.L.: A calcium-activated potassium channel present in foetal red cells of the sheep but absent from reticulocytes and mature red cells. Biochim. biophys. Acta *511:* 163–175 (1978).

Bygrave, F.L.: Mitochondrial calcium transport. Curr. Top. Bioenerg. *6:* 259–599 (1977).

Buckley, J.T.; Hawthorne, J.N.: Erythrocyte membrane polyphosphoinositide metabolism and the regulation of calcium binding: J. biol. Chem. *247:* 7218–7223 (1972).

Carafoli, E.: Plasma membrane Ca^{2+} transport and Ca^{2+} handling by intracellular stores: an integrated picture with emphasis on regulation by calcium; in Donowitz, Sharp, Mechanisms of intestinal electrolyte transport and regulation, pp. 121–134 (Liss, New York 1984).

Carafoli, E.; Crompton, M.: Calcium ions and mitochondria; in Duncan, Calcium in biological systems. Symposia of the Society for Experimental Biology, vol. 30, pp. 89–115 (Cambridge Univ. Press, Cambridge 1976).

Carafoli, E.; Crompton, M.: The regulation of intracellular calcium. Curr. Top. Membr. Transp. *10:* 151–216 (1978a).

Carafoli, E.; Crompton, M.: The regulation of intracellular calcium by mitochondria. Ann. N.Y. Acad. Sci. *307:* 269–284 (1978b).

Carafoli, E.; Lehninger, A.L.: A survey of the interaction of calcium ions with mitochondria from different tissues and species. Biochem. J. *122:* 681–690 (1971).

Carafoli, E.; Zuruni, M.: The Ca^{2+}-pumping ATPase of plasma membranes. Purification, reconstitution and properties. Biochim. biophys. Acta *683:* 279–301 (1982).

Caroni, P.; Reinlib, L.; Carafoli, E.: Charge movements during the Na^{2+}-Ca^{2+} exchange in heart sarcolemmal vesicles. Proc. natn. Acad. Sci. *77:* 6354–5358 (1980).

Chau-Wong, M.; Seeman, P.: The control of membrane bound Ca^{2+} by ATP. Biochim. biophys. Acta *241:* 473–482 (1971).

Cheung, W.Y.: Cyclic 3′,5′-nucleotide phosphodiesterase. Demonstration of an activator. Biochem. biophys. Res. Commun. *38:* 533–538 (1970).

Cheung, W.Y.: Calmodulin plays a pivotal role in cellular regulation. Science 207: 19–27 (1980).

Choquette, D.; Hakim, G.; Filoteo, A.G.; Plishker, G.A.; Bostwick, J.B.; Penniston, J.T.: Regulation of plasma membrane Ca^{2+}-ATPase by lipids of the phosphatidylinositol cycle. Biochem. biophys. Res. Commun. 125: 908–915 (1984).

Coffee, C.J.; Bradshaw, R.A.: Carp muscle calcium-binding protein. I. Characterization of the tryptic peptides and the complete amino sequence of component B. J. biol. Chem. 248: 3305–3312 (1973).

Cohen, P.; Cohen, P.T.; Shenolikar, S.; Nairn, A.; Victor, P.S.: The role of calmodulin in the structure and regulation of phosphorylase kinase of rabbit skeletal muscle. Eur. J. Biochem. 100: 329–337 (1979).

Crompton, M.; Capano, M.; Carafoli, E.: The sodium induced effluc of calcium from heart mitochondria. A possible mechanism for the regulation of mitochondrial calcium. Eur. J. Biochem. 69: 453–462 (1976).

DeLuca, H.F.; Engstrom, G.W.: Calcium uptake by rat kidney mitochondria. Proc. natn. Acad. Sci. USA 47: 1744–1747 (1961).

Downes, C.P.; Michell, R.H.: The control by Ca^{2+} of the polyphosphoinositide phosphodiesterase and the Ca^{2+}-pump in human erythrocyte. Biochem. J. 202: 53–58 (1982).

Dunham, E.T.; Glynn, I.M.: Adenosine triphosphatase activity and the active movements of alkali metal ions. J. Physiol., Lond. 156: 274–293 (1961).

Eaton, J.W.: The red cell; in Erythrocyte membranes: recent clinical and experimental advances, pp. 1–4 (Liss, New York 1981).

Ebashi, S.; Endo, M.: Calcium ion and muscle contraction. Prog. Biophys. molec. Biol. 18: 123–183 (1968).

Ebashi, S.; Ebashi, F.; Kodama, A.: Troponin as the Ca^{2+}-receptive protein in the contractile system. J. Biochem., Tokyo 62: 137–138 (1967).

Edmonson, J.W.; Li, T.K.: The effects of ionophore A23187 on erythrocytes. Relationship of ATP and 2,3-diphosphoglycerate to calcium-binding capacity. Biochim. biophys. Acta 433: 106–113 (1976).

Epel, D.E.; Wallace, R.W.; Cheug, W.Y.: Calmodulin activates NAD kinase of sea urchin eggs: an early event of fertilization. Cell 23: 543–549 (1981).

Evans, J.W.: Electrolyte concentrations in red blood cells of British breeds of sheep. Nature, Lond. 174: 931–932 (1954).

Evans, J.V.; Phillipson, A.T.: Electrolyte concentrations in the erythrocytes of the goat and ox. J. Physiol., Lond. 139: 87–96 (1957).

Fatt, P.; Ginsborg, B.: Measurement of calcium action potential in the crayfish. J. Physiol., Lond. 142: 516–543 (1958).

Ferreira, H.G.; Lew, V.L.: Ca transport and Ca pump reversal in human red blood cells. J. Physiol., Lond. 252: 86–87 (1975).

Gietzen, K.; Tejeka, M.; Wolf, H.U.: Calmodulin affinity chromatography yields a functional purified erythrocyte (Ca^{2+} + Mg^{2+})-dependent adenosine triphosphatase. Biochem. J. 189: 81–88 (1980).

Haiech, J.; Derancourt, J.; Pechere, J.F.; Demaille, J.G.: Magnesium and calcium binding to parvalbumins: evidence for differences between parvalbumins and an explanation of their relaxing function. Biochemistry 18: 2752–2758 (1979).

Hamill Q.P.; Marty, A.; Neher, E.; Sakmann, B.; Sigworth, J.F.: Improved patch-clamp techniques for high resolution current recording from cells and cell-free membrane patches. Arch. ges. Physiol. 391: 85–100 (1981).

Hasselbach, W.; Makinose, M.: Über den Mechanismus des Calciumtransportes durch die Membranen des sarkoplasmatischen Retikulums. Biochem. Z. *339:* 94–111 (1962).

Heilbrunn, L.V.; Wiercinski, F.J.: The action of various cations on muscle protoplasm. J. cell. comp. Physiol. *29:* 15–32 (1947).

Hoffman, J.F.: Cation transport and structure of the red cell plasma membrane. Circulation *26:* 1201–1213 (1962).

Horton, C.R.; Cole, W.Q.; Bader, H.: Depressed Ca²⁺ transport ATPase in cystic fibrosis erythrocytes. Biochem. biophys. Res. Commun. *40:* 505–509 (1970).

Jarrett, H.W.; Kyte, J.: Human erythrocyte calmodulin: further chemical characterization and the site of its interaction with the membrane. J. biol. Chem. *254:* 8237–8244 (1979).

Jenkins, R.E.; Lew, V.L.: Ca-uptake by ATP-depleted red cells from different species with or without associated increase in K⁺-permeability. J. Physiol., Lond. *234:* 41–42 (1973).

Kerr, S.E.: Studies on the inorganic composition of blood. IV. The relationship of potassium to the acid-soluble phosphorus fractions. J. biol. Chem. *117:* 227–235 (1937).

Klee, C.B.; Vanaman, T.C.: Calmodulin. Adv. Protein Chem. *35:* 213–321 (1982).

Kretsinger, R.H.: Hypothesis: Ca²⁺ modulated proteins contain EF hands; in Carafoli, Clementi, Drabikowski, Margreth, Calcium transport in contraction and secretion, pp. 468–478 (North Holland, Amsterdam 1975).

Kretsinger, R.H.: Structure and evolution of calcium-modulated proteins. CRC Crit. Rev. Biochem. *8:* 119–174 (1980).

Larsen, F.L.; Vincenzi, F.F.: Calcium transport across the plasma membrane: stimulation by calmodulin. Science *204:* 306–309 (1979).

Lehninger, A.L.; Carafoli, E.; Rossi, C.S.: Energy-linked ion movements in mitochondrial systems. Adv. Enzymol. *29:* 259–320 (1967).

Lehninger, A.L.; Veriesi, A.; Bababunmi, E.A.: Regulation of Ca²⁺ release from mitochondria by the oxidation-reduction state of pyridine nucleotides. Proc. natn. Acad. Sci. USA *75:* 1690–1694 (1978).

Le Maire, M.; Moller, J.V.; Tanford, C.: Retention of enzyme activity by detergent-solubilized sarcoplasmic reticulum Ca²⁺-ATPase. Biochemistry *15:* 2336–2342 (1976).

Linden, C.D.; Dedman, J.R.; Chafoleas, J.G.; Means, A.R.; Roth, T.F.: Interactions of calmodulin and coated vesicles from brain. Proc. natn. Acad. Sci. USA *78:* 308–312 (1981).

Locke, F.A.: The role of calcium in the contraction of isolated frog heart muscle. J. Physiol., Lond. *15:* 43–52 (1984).

MacLennan, D.H.: Purification and properties of adenosine triphosphatase from sarcoplasmic reticulum. J. biol. Chem. *245:* 4508–4518 (1970).

MacLennan, D.H.; Yip, C.C.; Iles, G.H.; Seeman, P.: Isolation of sarcoplasmic reticulum proteins. Cold Spring Harb. Symp. quant. Biol. *37:* 469–477 (1973).

Mac Manus, J.P.; Watson, D.R.; Yaguchi, M.: The complete amino acid sequence of oncomodulin – a parvalbumin-like calcium-binding protein from Morris hepatoma 5123tc. Eur. J. Biochem. *136:* 9–17 (1983).

Meissner, G.; Fleischer, S.: Characterization of sarcoplasmic reticulum from skeletal muscle. Biochim. biophys. Acta *241:* 356–378 (1971).

Meissner, G.; Conner, G.E.; Fleischer, S.: Isolation of sarcoplasmic reticulum by zonal centrifugation and purification of Ca^{2+} pump and Ca^{2+}-binding proteins. Biochim. biophys. Acta 298: 246–269 (1973).

Michell, R.H.: Two sites for Ca^{2+} control in one Ca^{2+} pump. Trends Biochem. Sci. 7: 123–124 (1982).

Niggli, V.; Penniston, J.T.; Carafoli, E.: Purification of the $(Ca^{2+} + Mg^{2+})$-ATPase from human erythrocyte membranes using calmodulin affinity column. J. biol. Chem. 254: 9955–9958 (1979b).

Niggli, V.; Ronner, P.; Carafoli, E.; Penniston, J.T.: Effects of calmodulin on the $(Ca^{2+} + Mg^{2+})$-ATPase partially purified from erythrocyte membranes. Archs Biochem. Biophys. 198: 124–130 (1979a).

Niggli, V.; Adunyah, E.S.; Penniston, J.T.; Carafoli, E.: Purified $(Ca^{2+} + Mg^{2+})$-ATPase of the erythrocyte membrane: reconstitution and effect of calmodulin and phospholipids. J. biol. Chem. 256: 395–401 (1981a).

Niggli, V.; Adunyah, E.S.; Carafoli, E.: Acidic phospholipids, unsaturated fatty acids, and limited proteolysis mimic the effect of calmodulin on the purified erythrocyte Ca^{2+}-ATPase. J. biol. Chem. 256: 8588–8592 (1981b).

Olorunsogo, O.O.; Okudolo, B.E.; Lawal, S.O.A.; Falase, A.O.: Erythrocyte membrane Ca^{2+}-pumping ATPase of hypertensive humans: reduced stimulation by calmodulin. Bioscience Reps. 5: 525–531 (1985).

Parker, J.C.: Interdependence of Ca and Na movements in dog red blood cells. Fed. Proc. 36: 271 (1977).

Parker, J.C.: Sodium and calcium movements in dog red blood cells. J. gen. Physiol. 71: 1–17 (1978).

Parker, J.C.: Active and passive Ca^{2+} movements in dog red blood cells and resealed ghosts. Am. J. Physiol. C 237: C10–C16 (1979).

Parker, J.C.; Gitelman, H.J.; Glosson, P.S.; Leonard, D.L.: Role of calcium in volume regulation by dog red blood cells. J. gen. Physiol. 65: 84–96 (1975).

Penniston, J.T.: Plasma membrane Ca^{2+}-pumping ATPases. Ann. N.Y. Acad. Sci. 402: 296–303 (1982).

Penniston, J.T.: Plasma membrane Ca^{2+}-ATPases as active Ca^{2+} pumps; in Cheung, Calcium and cell function, vol. 4, pp. 99–149 (Academic Press, New York 1983).

Perry, S.V.: The regulation of contractile activity in muscle. Trans. biochem. Soc. 7: 593–617 (1979).

Porzig, H.: ATP-independent calcium net movements in human red cell ghosts. J. Membr. Biol. 8: 237–258 (1972).

Peterson, S.C.; Kirshner, L.B.: Di- and triphosphoinositide metabolism in intact swine erythrocytes. Biochim. biophys. Acta 202: 283–294 (1970).

Peterson, S.W; Ronner, P.; Carafoli, E.: Partial purification and reconstitution of the $(Ca^{2+} + Mg^{2+})$-ATPase of erythrocyte membranes. Archs Biochem. Biophys. 186: 202–210 (1978).

Quist, E.E.; Roufogalis, B.D.: Calcium transport in human erythrocytes, separation and r-constitution of the high and low affinity prepared at low ionic strength. Archs Biochem. Biophys. 168: 240–251 (1975).

Rasmussen, H.: Cell communication calcium ion and cyclic adenosine monophosphate. Science 170: 405–412 (1970).

Rasmussen, H. and Goodman, D.B.P.: Relationships between calcium and cyclic nucleotides in cell activation. Physiol. Rev. 57: 421–509 (1977).

Reeves, J.P.; Sutko, J.L.: Sodium-calcium exchange in cardiac membrane vesicles. Proc. natn. Acad. Sci. USA 76: 590–594 (1979).

Reuter, H.: Ion channels in cardiac cell membranes. Annu. Rev. Physiol. 46: 473–484 (1984).

Reuter, H.; Seitz, N.: The dependence of Ca²⁺ effluc from cardiac muscle on temperature and external ion composition. J. Physiol., Lond. 195: 451–470 (1968).

Reuter, H.; Stevens, C.F.; Tsien, R.W.; Yellen, G.: Properties of single calcium channels in cardiac cell culture. Nature, Lond. 297: 501–504 (1982).

Ringer, S.: Regarding the action of hydrate of soda, hydrate of ammonia, and hydrate of potash on the ventricles of the frog's heart. J. Physiol., Lond. 3: 195–202 (1882).

Ringer, S.: A further contribution regarding the influence of the different constituents of the blood on the contraction of the heart. J. Physiol., Lond. 4: 29–42 (1883).

Robinson, G.A.; Butcher, R.W.; Sutherland, E.W.: Cyclic AMP, pp. 1–47 (Academic Press, New York 1971).

Roelofsen, B.; Schatzmann, H.J.: The lipid requirement of the Ca²⁺, Mg²⁺-ATPase in the human erythrocyte membrane as studied by various highly purified phospholipases. Biochim. biophys. Acta 464: 17–36 (1977).

Ronner, P.; Gazzotti, P.; Carafoli, E.: A lipid requirement for the Ca²⁺, Mg²⁺-activated ATPase of erythrocyte membranes. Archs Biochem. Biophys. 179: 578–583 (1977).

Rottenberg, H.; Scarpa, A.: Calcium uptake and membrane potential in mitochondria. Biochemistry 13: 4811–4819 (1974).

Roufogalis, B.D.: Regulation of calcium translocation across the red blood cell membrane. Can. J. Physiol. Pharmacol. 57: 1331–1349 (1979).

Scharff, O.: The influence of calcium ions on the preparation of the (Ca²⁺ + Mg²⁺)-activated membrane ATPase in human red cells. Scand. J. clin. Lab. Invest. 30: 313–320 (1972).

Schatzmann, H.J.: ATP-dependent Ca²⁺-extrusion from human red cells. Experientia 22: 364–365 (1966).

Schatzmann, H.J.: Correlations between (Na⁺ + K⁺)-ATPase, Ca²⁺-ATPase and cellular potassium concentrations in cattle red cells. Nature, Lond. 248: 58–60 (1974).

Schatzmann, H.J.: The plasma membrane calcium pump of erythrocytes and other animal cells; in Carafoli, Membrane transport of calcium, pp. 41–108 (Academic Press, London 1982).

Schatzmann, H.J.; Rossi, G.L.: (Ca²⁺ + Mg²⁺)-ATPase in human red cells and their possible relations to cation transport. Biochim. biophys. Acta 241: 379–392 (1971).

Schatzmann, H.J.; Scheidegger, H.R.: Postnatal decline of (Ca²⁺ + Mg²⁺)-activated membrane ATPase in cattle red cells. Experientia 31: 1260–1261 (1975).

Schatzmann, H.J.; Vincenzi, F.F.: Calcium movements across the membrane of human red cells. J. Physiol., Lond. 369–395 (1969).

Seeman, K.B.; Kretsinger, R.H.: Calcium modulated proteins; in Spiro, Metal ions in biology, vol. 6 (Academic Press, New York 1983).

Sheetz, M.P.; Febbroriello, P.; Koppel, D.: Triphosphoinositide increases glycoprotein lateral mobility in erythrocyte membranes. Nature, Lond. 296: 91–93 (1982).

Stieger, J.; Schatzmann, H.J.: Metal requirement of the isolated red cell Ca-pump ATPase after elimination of calmodulin dependence by trypsin attack. Cell Calcium 2: 601–616 (1981).

Sutherland, E.W.; Robinson, G.A.; Butcher, R.W.: Some aspects of the biological role of adenosine 3′,5′-monophosphate (cAMP). Circulation *37:* 279–306 (1968).

Szebenyi, D.M.E.; Obendorf, S.K.; Moffat, K.: Structure of vitamin D dependent calcium-binding protein from bovine intestine. Nature, Lond. *294:* 327–332 (1981).

Tada, M.; Kirchberger, M.; Li, H.C.: Phospholambanphosphatase-catalyzed dephosphorylation of the 22,000 dalton phosphoprotein of cardiac sarcoplasmic reticulum. J. Cyclic Nucl. Res. *1:* 329–338 (1975).

Tada, M.; Yamamoto, T.; Tonomura, Y.: Molecular mechanism of active calcium transport by sarcoplasmic reticulum. Physiol. Rev. *58:* 1–79 (1978).

Tanaka, T.; Ohmura, T.; Yamakado, T.; Hidaka, H.: Two types of calcium-dependent protein phosphorylations modulated by calmodulin antagonists: naphthalenesulfonamide derivatives. Molec. Pharmacol. *22:* 408–412 (1982).

Tash, J.S.; Mann, A.R.: Regulation of protein phosphorylation and motility of sperm by cyclic adenosine monophosphate and calcium. Biol. Reprod. *26:* 745–763 (1982).

Taverna, R.D.; Hanahan, D.J.: Modulation of human erythrocyte Ca^{2+}, Mg^{2+}-ATPase activity by phospholipase A_2 and proteases: a comparison with calmodulin. Biochem. biophys. Res. Commun. *94:* 652–659 (1980).

Ting, A.; Lee, J.W.; Vidaver, C.A.: Calcium transport by pigeon erythrocyte membrane vesicles. Biochim. biophys. Acta *555:* 239–248 (1979).

Tosteson, D.C.: Active transport genetics and cellular evolution. Fed. Proc. *22:* 19–26 (1963).

Tosteson, D.C.; Hofmann, J.F.: Regulation of cell volume by active cation transport in high and low potassium sheep red cells. J. gen. Physiol. *44:* 169–194 (1960).

Vasington, F.D.; Murphy, J.V.: Active binding of calcium by mitochondria. Fed. Proc. *20:* 146 (1961).

Vasington, F.D.; Murphy, J.V.: Ca^{2+} uptake by rat kidney mitochondria and its dependence on respiration and phosphorylation. J. biol. Chem. *237:* 2670–2672 (1962).

Verecka, L.; Carafoli, E.: Vanadate induced movements of Ca^{2+} and K^+ in human red blood cells. J. biol. Chem. *257:* 7414–7421 (1982).

Vianna, A.L.: Interaction of calcium and magnesium in activating and inhibiting the nucleoside triphosphatase of sarcoplasmic reticulum vesicles. Biochim. biophys. Acta *410:* 389–406 (1975).

Walsh, M.P.; Cavadore, J.C.; Vallet, B.; Demaille, J.G.: Calmodulin-dependent myosin light chain kinase from cardiac and smooth muscle: a comparative study. Can. J. Biochem. *58:* 299–308 (1980).

Welsh, M.J.; Dedman, J.R.; Brinkley, B.R.; Means, A.R.: Calcium-dependent regulator protein: localization in mitotic apparatus of eukaryotic cells. Proc. natn. Acad. Sci. USA *75:* 1867–1877 (1978).

Wikstrom, M.; Ahonen, P.; Luukkainen, T.: The role of mitochondria in uterine contraction. FEBS Lett. *56:* 120–123 (1975).

Wong, P.Y.K.; Cheung, W.Y.: Calmodulin stimulates human platelet phospholipase A_2. Biochem. biophys. Res. Commun. *90:* 473–480 (1979).

Enitan A. Bababunmi, PhD, DSci, Biomembrane Research Laboratories, Department of Biochemistry, College of Medicine, University of Ibadan, Ibadan (Nigeria)

Simopoulos AP (ed): Selected Vitamins, Minerals, and Functional Consequences of Maternal Malnutrition. World Rev Nutr Diet. Basel, Karger, 1991, vol 64, pp 139–173

Functional Consequences of Maternal Malnutrition

Teresa González-Cossío, Hernán Delgado [1]

Division of Nutrition and Health, Institute of Nutrition of Central America and Panamá, Guatemala City, Guatemala

Contents

[1] We would like to thank the generous technical support of Dr. W. Bruce Currie of the Cornell. University Animal Science Department, and the secretarial help of Miss Diana Cofiño Palala and Gladys de González from INCAP.

I. Introduction

The study of maternal malnutrition has a several-fold importance. Malnutrition is a social condition undesirable for both sexes and across ages and sexes. It is a sensitive indicator of poverty and, independently of its functional significance, it is an unacceptable human condition.

The functional consequences of malnutrition are currently being explored and better understood. There is an association between malnutrition and some undesirable functional outcomes. A considerable body of evidence has shown that immunodepression is experienced by the malnourished young as well as by the adult, accompanied by longer episodes of illness [Black et al., 1984; Cunningham-Rundles, 1982; Chandra, 1981]. Malnourished people have a reduced work capacity [WHO, 1979], and malnutrition has been associated with other adverse outcomes such as impaired psychomotor development [Chávez et al., 1975b; Cravioto and DeLicardie, 1970] and an increased incidence of mortality [Chen et al., 1980]. It is now accepted that malnutrition might not be the only causal factor of such outcomes, but rather that it coexists in an environment that does not promote an adequate psychomotor development [Cravioto and DeLicardie, 1975] and where overcrowding and poor sanitary conditions favor the development of diseases. These conditions, coupled with limited access to medical care, are the link to the excess morbidity and mortality observed in the poor segments of populations.

But maternal malnutrition has much broader functional consequences. Malnutrition during childhood and puberty causes short stature and depleted reserves of some nutrients such as iron. Short stature is an important nutritional indicator that is independently associated with intrauterine growth retardation (IUGR) [Anderson et al., 1984; Dougherty and Jones, 1982]. When pregnancy occurs, maternal nutritional demands are increased and in malnourished women, these are rarely met. Low weight gain will occur if intake is inadequate and the probability of delivering a low birth weight (LBW) infant is substantially increased under these conditions [Winikoff and Debrovner, 1981; Metcoff et al., 1981]. Depending on the timing, the type and the magnitude of the nutritional insult the mother is exposed to during pregnancy, the IUGR neonate may have either an adequate or an inadequate ponderal index [Villar and Belizán, 1982]. During lactation the nutrient requirements are increased compared to those of pregnancy. Women consuming a poor diet have been observed to produce insufficient amounts of milk [Hanafy et al., 1972] which is associated with poor infant growth [Chávez et al. 1975a; Waterlow and Thompson, 1979; Jelliffe and Jelliffe, 1979].

In summary, malnourished women are not only likely to suffer from depressed immune response and exhibit reduced work capacity, they also have an increased probability of delivering a LBW infant whom they are unlikely to feed adequately early in life. The implications of malnutrition in terms of public health are clearly important for the well-being, health and productivity of populations.

It is therefore important to understand the mechanisms underlying these conditions so that public health interventions to address these problems can be designed adequately. Some of the most well known functional consequences of maternal malnutrition are LBW and poor lactation performance.

There are several areas of interest related to these phenomena. These include:

(1) Maternal diet. An IUGR infant can be the product merely of a limited delivery of nutrients to the feto-placental unit during gestation because of a deficient maternal diet. The supply can be limited in energy and/or in certain specific nutrients such as proteins or iron.

(2) Placental factors. Retarded growth during intrauterine life may reflect a functionally impaired placenta. Reductions in placental size may impair overall transfer between the mother and the conceptus. Functional placental capacity can be reduced due to infarcts or because of reduced villous or microvillous area, either of which could reduce the capacity for nutrient transfer to the fetus, or the hormonal delivery to the mother. Placental metabolic changes might comprise reductions of nucleic acids, alterations in the activity of enzymes and synthesis of hormones that regulate the course of pregnancy.

(3) Hemodynamic factors. Malnutrition in utero might result from alterations of the hemodynamic changes that normally take place during gestation. Possible alterations are inadequate blood volume expansion, decreased blood pressure or inappropriate distribution of cardiac output. These processes would manifest themselves in inadequate uterine blood flow, thereby limiting nutrient supply to the conceptus.

(4) Other non-nutritional factors associated with LBW. A number of placental infections are present in a high proportion of malnourished women. Some placental infections damage its structure and interfere with an adequate transfer of nutrients and fetal wastes across the placenta. Adolescence is also associated with LBW. A pregnant child who is still growing has

larger nutritional requirements than a mature pregnant woman. Also, ado-
lescents often have inadequate food habits, and these factors help to explain
the excess prevalence of LBW observed in this age. The physical activity of
the pregnant women is also related to LBW. IUGR might be linked to
strenuous physical activity through competition for available energy in
which case growth retardation would be secondary to an inadequate nutrient
supply. Another possibility is that vigorous physical activity, or a specific
position adopted during long periods, may interfere with normal uterine
blood flow preventing a normal delivery of nutrients to the conceptus.

(5) Poor lactation performance. The amount of milk a lactating
woman produces is related to a number of important factors, one of which
is the size of the baby. In this way, LBW may be the mechanism that
explains why undernourished women produce inadequate amounts of
milk. However, LBW might not be the only explanation of these observations. A reduced milk output has been repeatedly observed in poorly nourished women [Jelliffe and Jelliffe, 1979]. Even though, the relative importance of maternal malnutrition and of breast feeding patterns have not
been sorted out unambiguously, there is recent evidence suggesting that
poor maternal energy reserves may compromise lactation performance
[Brown et al., 1986].

There are sufficient human and animal studies in the above described
areas to discuss the functional consequences of maternal undernutrition
during the reproductive years.

The purpose of this paper is to offer a review of what functions are
altered in malnourished pregnant and lactating women. This will consist of
summarizing the nutritional determinants of LBW and a discussion of
some of the biological mechanisms through which these determinants
might be operating. Finally, the impact of undernutrition during lactation
will be discussed.

II. Nutritional Determinants of Birth Weight

A. Birth Weight Characterization
Birth weight (BW) is determined by the gestational age and the growth
status of the fetus. When the determinants of LBW are studied, a distinction between prematurity and IUGR must be made.

A further classification of IUGR neonates is needed. Fetal malnutrition is the result of a nutrition insult during pregnancy. The final effect on

fetal growth depends on the magnitude an duration of the insult and, very importantly, on the growth stage at which this insult occurs. Thus, if the maternal nutritional deprivation sets on early in pregnancy, when peak height gain velocity occurs [Tanner, 1978], fetal linear growth will be compromised. Conversely, if the insult occurs only late in pregnancy, when peak weight gain velocity is found, weight accretion will be limited. The latter newborn has an adequate length at birth, but a low weight for length, which means having a low ponderal index (LPI). The former neonate will have adequate weight for length, or an adequate ponderal index (API) [Villar and Belizán, 1982; Tanner, 1978]. Naturally the magnitude and timing of these insults may vary, with fetal growth varying accordingly.

Prematurity and, as just discribed, IUGR-LPI or IUGR-API have different etiologies and different distributions between countries. This is an important concept in nutritional planning because each type of LBW calls for a distinct intervention. The impact of these interventions on the population will depend on the relative proportions of each group.

The diverse groups of LBW neonates also have different prognosis in terms of mortality. High early mortality is the clearest functional consequence known of LBW. Neonatal mortality sharply raises below 2,500 g at birth, even when no distinction about the type of LBW is made (for a good example, see Chase [1969]). However. different types of LBW infants experience clearly distinct neonatal mortality rates [Haas et al., 1987]. Infant mortality, morbidity, physical growth, and neurological development are affected by LBW, and are also differentially impaired amongst the LBW groups [Ferguson, 1978; McCormick, 1985; Villar et al., 1984; Ashworth and Feachem, 1985].

B. Maternal Nutritional Status and Birth Weight

1. Introduction

A vast number of factors existent before and during pregnancy have an influence on the weight at birth of the neonate. These include behavioral factors such as maternal smoking or strenuous exercise during pregnancy; both of which, depending on the degree of exposure, have a strong negative influence on BW [Martorell and González-Cassío, 1987].

Other factors include biodemographic characteristics of the mother such as age, parity, marital status or pregnancy interval; health status such as infections or chronic diseases, and socioeconomic variables such as maternal education, family income or occupation of the head of the household.

All these non-nutritional variables relate directly (biologically) or indirectly with weight at birth. Nutritional status of the mother, both before and during pregnancy, is also related to the weight of the newborn. This means that BW is sensitive but not specific to maternal nutritional status. In fact, the sensitivity of BW to variations of the mother's nutritional status is rather modest, compared to that of other conditions such as hypertension or tobaccoism. However, there is such a large number of undernourished women in the world, that the attributable proportion of LBW to maternal malnutrition is large and important in terms of public health.

The impact of maternal nutritional status on BW depends on the type, magnitude and timing of the deficiency.

Randomized therapeutic experimental trials are adequate procedures to study the independent effect of specific nutrient deficiencies on BW. The validity of the findings of such studies would depend, among other things, on the recipients being truly needy. A lack of effect of a supplementation on BW in a nutrient-sufficient population offers no information on the possible impact of such supplementation in a truly needy group. The lack of necessity of a supplementation or the irrelevant comparisons made between nutritional sufficient populations are perhaps the strongest limitations of the studies in this area. Many intervention trials have been done in not needy populations, and good examples of this issue are reviewed by Hemminki and Starfield [1978a, 1978b] and by Kramer [1985].

Detailed analysis of the impact of deficiencies in all vitamins and minerals on BW is out of the scope of this document. (The interested reader is referred to the three revisions just mentioned.)

Briefly, the impact on BW of vitamin B_{12}, folate, calcium and iron nutriture may be important in nutrient deficient populations. The relative relevance of the nutritional status of these elements remains to be defined in needy groups.

The relationship of BW with some anthropometric and dietary indicatos of maternal nutritional status have been studied in well-fed and undernourished mothers. These indicators are maternal height, prepregnant weight, pregnancy weight gain and caloric/protein intake (the former as a proxy for energy balance). In order to understand the independent effect of these indicators on BW, control of potential confounding factors must be achieved, be it by design or by analysis. This concept is important because all the indicators listed above are interrelated, and if potential confounding variables are not adequately taken into account, erroneous conclusions may be arrived at.

2. Maternal Height

The stature of the mother during pregnancy has been repeatedly observed to be independently associated with intrauterine growth [Hytten and Chamberlain, 1980; Habicht et al., 1973; Lechtig et al., 1975 b, c] and more recent evidence suggests that there is also an association with gestational age [Delgado et al., 1982]. Stature effects are independent of weight at the beginning of pregnancy and of weight gain during the same period. Maternal stature is an indicator of her prior nutritional status. High prevalences of low height in a population are indicative of undernutrition during periods of growth. This measurement may be used to identify women at risk of delivering LBW infants [Lechtig et al., 1976]. The cut-off point used for such purpose will depend on the distribution of maternal stature and on the prevalence of low maternal stature in the population. It should be kept in mind that the effect of stature on BW can not be overcome by interventions during adulthood.

3. Prepregnant Weight

Low weight at the beginning of pregnancy, controlling for stature, has an independent negative influence on the growth of the fetus as well as on the gestational age [Kaminski et al., 1973; Edwards et al., 1979; Anderson et al., 1984]. Even when adequate weight increments during pregnancy are observed, the effect of low pregravid weight on BW is observed [Edwards et al., 1979]. This concept has twofold importance. First, maternal prepregnant weight, as in the case of stature, can be used to identify women at risk of deliveries a LBW infant. Here again, the selection of the cut off point will depend on the distribution of weight in the population and on the prevalence of low prepregnant weight, as well as on the available resources for treatment. Second, of probably more public health importance is that low weight (for height) as opposed to low height is a treatable type of adult undernutrition, with a potential to lower the incidence of LBW in malnourished populations.

4. Pregnancy Weight Gain

There is a straight forward positive relationship between maternal weight gain during pregnancy and BW. Fetal BW represents approximately

¼–⅓ of the total pregnancy weight gain. The rest is accounted for by placenta, amniotic fluid, maternal reserves such as fat and protein in uterus and mammary glands, and intra- and extravascular fluid increments [Hytten and Chamberlain, 1980].

The association between pregnancy weight increments and BW is present even after accounting for prepregnancy weight, height, and gestational age [Anderson et al. 1984]. Lechtig and Klein [1980b] reported high incidences of LBW neonates born to mothers experiencing low pregnancy weight gains, especially in those women who were underweight (less than 46 kg) at the beginning of pregnancy. These figures show that 60% of the newborns had LBW among underweight mothers who gained less than 3.5 kg at term. Incidence of LBW rapidly decreases as total weight gain increases in these women. The same strong effect of pregnancy weight gain on BW as in underweight women is not observed in women with high prepregnancy weight. Normal and overweight women have lower rates of LBW and they would be expected to show diminished decrements in LBW rates as weight gain increases. This is to be expected if prepregnancy weight reflects nutrient reserves for the growing fetus, such that when these are high, the fetus does not have to rely so much on events during pregnancy.

Several studies have observed the effect modification (biological interaction) that prepregnancy weight has on the relationship between weight gain and BW [Winikoff and Debrovner, 1981; Naeye, 1981a, 1981b].

These studies indeed show a smaller benefit of the weight gained in pregnancy in terms of BW as prepregnancy weight increased.

5. Maternal Diet

It was previously mentioned that mineral and vitamin nutritional status of pregnant women have probably a small impact on the weight of the newborn. Protein and calories may have an important influence on BW for these diet constituents are the inputs to synthetize new tissue. The relative importance of each may depend on the usual intake of the mother. Proteins may be more important than calories in populations with caloric-sufficient/protein-deficient diets, as in diets where cassava is the most important staple. Similarly calories in energetically poor diets. But if caloric intake is low, proteins will be used as a source of calories. So a caloric deficient diet may render an otherwise protein adequate diet inadequate.

Supplementation studies during pregnancy offer an opportunity to evaluate this issue.

Reviewing the literature, it can be observed that the usual home intake of the recipient women in supplementation studies was generally low in caloric content (1,200–1,950 kcal/day excluding the New York group), but adequate in protein intake for nonpregnant women, both in terms of grams per kilogram of body weight, and in terms of the percent of total calories that is derived from proteins. The protein intake of the women in supplementation studies ranged from 35 to 45 g protein/day which should cover 95% of the protein requirements of nonpregnant women weighing 45–50 kg. This requirement increases 4–10 g/day during pregnancy [NAS, 1980; Mora et al., 1979; Lechtig et al., 1975a; Iyengar and Rajalakshmi, 1975; Chávez and Martínez, 1979; Rush et al., 1980; McDonald et al., 1981]. It is thus hard to estimate the magnitude of the relationship between BW and varying intakes of proteins because there has not been enough variability in protein intake in the studies reviewed.

The evidence for calories is clearer. Supplementation studies concerning this show that maternal diet is related to the weight of the newborn. More interesting, the results show that maternal nutritional status modifies this relationship. A given caloric intake has a larger impact on BW on smaller or more undernourished women than on the better-nourished women [Lechtig et al., 1975a]. The effect ranges from 28 to 71 g of BW increment per 10,000 net supplemented kcal during pregnancy. The latter figure was observed in a group of Indian women pertaining to a population with high prevalences of maternal malnutrition [Iyengar and Rajalakshmi, 1975], and the relationship of 28 g of BW per 10,000 net supplemented kcal was observed in a group of women at risk of delivering LBW infants in New York [Rush et al., 1980]. The nutritional status of the recipient women of this study was the best reported in all the supplementation studies during pregnancy.

6. Conclusions

The role of maternal malnutrition has been described in explaining the high incidence of LBW in developing countries. The relative importance depends on the prevalence of each condition in the population studied. We can use data of prevalence of some maternal characteristics to exemplify the strong public health importance of maternal malnutrition. The etiologic fraction and the relative risk at given cut-off points of the variables of interest can be used as the basis for this evaluation. Prevalence data were obtained from Guatemala [MSPAS and INCAP, 1986] and values for the relative risks from Kramer [1985]. Results are presented in table 1.

Table 1. Relative importance of some LBW determinants in a developing country

Determinant	Prevalence %	Relative risk[1]	EF[2], %	Sensitive to food supplementation
Low weight gain or low caloric intake in pregnancy	59[3]	1.98[4]	39.6	yes
Low prepregnancy weight	50[5]	1.84	29.6	yes
Low stature	86.8[6]	1.27	19.0	no
Race (non-Caucasian)	50[7]	1.39	16.3	no
Sex of neonate (female)	51.3[8]	1.19	8.9	no
Primiparity	12.3[9]	1.23	2.7	no
Malaria during pregnancy	?[10]	–	–	–
Obstetric history (previous LBW, abortions)	?	–	–	–
Weight and height of father	?	–	–	–

[1] Relative Risk (RR) refers to IUGR and not prematurity. RR of IUGR is close to the RR LBW in developing countries where excess LBW is explained mostly by IUGR and not prematurity.

[2] $EF = \text{etiologic fraction} = \dfrac{\text{prevalence}\,(RR - 1)}{\text{prevalence}\,(RR - 1) + 1}$

or the fraction attributed to a given variable.

[3] 59% of women gain less than 7 kg during pregnancy in Guatemala (data from Lechtig and Klein [1980b], and assuming a normal distribution of weight gain). 7 kg was the cut-off point used to study well nourished women. RR is greater in malnourished women due to the interaction between prepregnant weight and weight gain.

[4] RR from Kramer [1985]. The cut-off points are indicated in footnotes to each determinant.

[5] Mean weight of nonpregnant nonlactating women aged 15–49 years living in the North-Guatemala health area of Guatemala City is 50.3 kg. 50 kg was the cut-off point used.

[6] 86.8% of the women described in footnote 5 had a height < 1.57 m. This was the cut-off point used.

[7] In Guatemala approximately 50% of the women are ladinas (non-Indian). This is a cultural characterization but was the best available approximation.

[8] 51.3% males is the normal sex ratio at birth. Inadequate perinatal conditions or high prevalence of LBW decrease this proportion [Weller and Bouvier, 1980]. In Guatemala, average BW is 3.2 kg and sex ratio at birth would be approximately 48.3%.

[9] In the North-Guatemala City health area, 12.3% of women aged 15–49 years have only one child.

[10] No estimates of prevalence or RR for the rest of the determinants were found.

From exercises like this it can be shown that in a developing country maternal malnutrition explains the greatest proportion of LBW, and that caloric malnutrition may account for up to $^2/_3$ of the incidence of IUGR. This type of malnutrition is preventable by interventions aimed at improving maternal intake during pregnancy, as has been shown by experimental supplementation trials conducted in such countries as Guatemala or Gambia [Lechtig et al., 1975a; Prentice et al., 1983].

C. Placental Role in Low Birth Weight
The physiology of the placenta and its alterations during undernutrition will be discussed in this section.

1. Physiology of the Placenta
After the ovum is fertilized it travels through the ovarian tube and buries itself into the endometrium. Proliferating embryonal and endometrial tissues start forming the placenta. This fetal organ has several functions: transfer of nutrients to the fetus and excretion of fetal waste into the maternal circulation; protection of the fetus from environmental hazards including the maternal immune mechanisms; endocrine secretions to control the course of pregnancy and modification of the maternal metabolism, placenta's own metabolism, synthesis of glycogen, fatty acids and cholesterol which are critical substrates for normal placental function [Moore, 1977; Gruenwald, 1975].

a) Placental Characteristics. Placental transfer function depends on several factors such as the surface area available for exchange, the relative concentration of nutrients and oxygen in the blood, the uteroplacental blood flow, the thickness and morphology of the membranes through which transfer occurs, and the transfer mechanisms for individual substances [Morris, 1981; Fox, 1979; Munro, 1980].

The placenta is a fetal organ that has many villi containing the fetal vessels. The outer part of the villi is the trophoblast (syncytial and cytotrophoblast) which is the metabolic component of the placenta. The trophoblast, the connective tissue and the endothelium of the fetal capillaries constitute the placental membranes which undergo changes throughout the pregnancy. The villi branch and mature becoming larger in number and smaller in size. Fetal vessels are sinusoidally dilated to occupy most of the villous cross-section and are peripherally placed in apposition to an increasingly thinned trophoblast. In the second part of pregnancy, the cytotrophoblast no longer forms a continuous layer, thus contributing to the thinning of the membrane. Total villous surface area increases from 5 to 11 m^2 by term [Moore, 1977; Fox, 1979]. The thickness of the syncytiotrophoblast decreases considerably from 10 μm early in pregnancy to approximately 1.7 μm at term. All these changes enormously increase the placental transfer capacity.

Normal-term placentas have immature villi which are supposed to be new branches evidencing its constant growth capacity. This is also perceived in adverse circumstances in which placental mass increase beyond normal size, as in decompensated heart disease, severe anemia, or high altitude [Fox, 1979]. The placenta has also a considerable functional reserve. Good evidence for this reserve is given by studies of obstruction of some transfer sites of the placenta. Usually, there is perivillus fibrin deposition throughout pregnancy which is a consequence of the normal wear and tear, and it is not associated with IUGR or hypoxia unless the plaques cover more than 30% of the total placental surface. Certain amount of thrombi are also unrelated to IUGR, unless they are secondary to high blood pressure, when a restriction of blood flow to the conceptus occurs [Fox, 1979].

Given that the placenta has a constant growth capacity and an important functional reserve, it would be incorrect to infer automatically that a small placenta attached to a IUGR neonate is the cause of the fetal growth retardation.

Small placentas should be closely analysed for functional problems, such as inadequate transfer capacity or endocrine production, before causality can be inferred.

As has been argued [Gruenwald, 1975; Fox, 1979] to assume that a small placenta is the cause of IUGR requires the further assumption that the placenta functions at its maximal capacity under normal circumstances (a necessary condition to limit fetal growth) and, based on the evidence presented above, this is now accepted not to be the case.

A small placenta may well be the product of a small fetus, the smallness of which caused by extraplacental factors, with the (small) placenta functioning normally. Consequently, in order to study the role of the placenta in IUGR, it is necessary to study not only gross placental characteristics such as weight or volume, but also to study surface area of villi and distances between relevant tissue compartments, as well as trophoblastic characteristics, in order to draw conclusions about placental functionality.

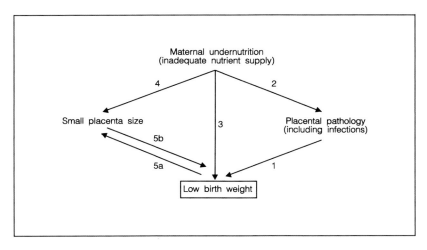

Fig. 1. Relationships between maternal undernutrition and LBW.

2. Placental Alterations during Undernutrition

The relationships between maternal undernutrition and LBW can be described using the model presented in figure 1. Maternal undernutrition, measured by anthropometry, is used as a proxy for inadequate nutrient supply. This model proposes the mechanisms by which LBW can occur. Some of these associations are biological and the direction suggests causality. Others are only statistical, meaning that these events happen simultaneously but that there is no clear evidence in the literature to support the causality of these relationships.

The role of an infected or otherwise physically damaged placenta in the etiology of LBW is relatively clear (link 1 in figure 1). It seems, for example, that certain bacterial infections during pregnancy such as *Ureaplasma urealyticum* or *Chlamydia trachomatis* are independently associated with IUGR [Kramer, 1985].

Thorough analysis of the relationship between maternal undernutrition and infection (including placental) (link 2) is out of the scope of this paper. Briefly, there is evidence that moderate undernutrition in children impairs cell-mediated immune response (CMIR) and severe undernutrition impairs CMIR in children as well as in adults [Rivera, 1984; Black et al., 1984; Chandra, 1981]. Undernutrition prevails in unfavorable environments where infectious diseases are common; and whether moderate maternal undernutrition (commonly observed in women delivering LBW infants in developing countries) is the cause of increased placenta infections, or if these events only coincide, remains to be studied.

There is little doubt now that maternal undernutrition is an independent factor in the etiology of IUGR (link 3). As discussed above, the indicators of undernutrition such as low height, low prepregnant weight, low weight gain, and inadequate dietary intake during pregnancy have been shown to be important predictors of LBW [Kramer, 1987; Institute of Medicine, 1985].

There is a well known and biologically sound relationship between placental weight and BW (link 5b) [Hytten and Chamberlain, 1980; Kaiser, 1977; Bonds et al., 1984].

By virtue of the coexistence of the biological relationships 3 and 5b, link 4 is correspondingly often observed [Lage et al., 1972; Lechtig et al., 1975c; Murthy et al., 1967; Dickin, 1986].

In this scenario, LBW could be the reflection of inadequate nutrient supply to the conceptus (link 3). A small fetus does not need a large placenta, and the small placenta will then be a reflection, rather than the cause, of LBW. This can be described by link 3 occurring first and followed by link 5a.

If this last relationship is described with a regression equation, BW = mean BW + B1 (inadequate supply) + B2 (placental characteristics), the last statement presented would mean that B2 (placental characteristics) would not be an important predictor of LBW when used in the model after inadequate supply. However, there is a fundamental problem when using only a statistical approach to answer this question. Placental size, even when not playing an independent role in the etiology of LBW, is a variable that is biologically closer to LBW than variables related to maternal nutritional status (such as weight for height). This means that the statistical association will be stronger for placental than for other maternal characteristics even if placenta does not play an independent role (this holds if we assume that measurement errors of both variables are similar).

Using the statistical approach coupled with the study of the micromorphometry of the placentas attached to IUGR infants is necessary to clarify this research question. This issue has practical implications. If maternal undernutrition at a certain time during pregnancy affects placental development in such a way as to interfere with its later function, then efforts to overcome maternal malnutrition after this insult might not be as efficient as if a functional insult to the placenta might had not taken place.

Although there is considerable research in this area, an agreement has not been reached as yet. This last point is exemplified by the following discussion of research addressing these issues, mostly in humans.

Several studies have been undertaken to analyse the influence of maternal undernutrition on the placenta. Laga et al. [1972] analysed placentas of two different groups of women, one from the United States and the other from Guatemala. Women were not comparable in nutritional status or socioeconomic characteristics, the Guatemalans being at a disadvantage.

Placentas of the Guaemalan women were smaller by approximately 65 g (uncontrolled for gestational age), had more fibrin deposition and more evidence of chronic infection. They also had less peripheral villous mass and surface area as well as less trophoblastic mass. Stem villi and surface area were not significantly different between the groups. Villous cross-section analysis failed to reveal differences between placentas of these groups, suggesting that there is not individual villous change but rather a difference in the number of them. When placentas were analysed controlling for birth weight and gestational age, the Guatemalan samples showed an especially greater deficit in peripheral villous and trophoblastic mass and in peripheral villous surface. Interestingly, these results, rather than demonstrating inadequate Guatemalan placentas, they do the opposite, i.e., they seem to show that even when there were differences in the parenchyma, these were unrelated to birth weight, because deficits were still present after controlling for birth weight and gestational age. Thus, differences in the amount of relevant tissue for diffusion of nutrients, placental metabolism and endocrine production were not in themselves limiting fetal growth. Probably, in this study LBW was due to extra-placental factors.

Some time later, Lechtig et al. [1975c] in rural and urban Guatemala studied the association between maternal undernutrition and placental characteristics. There were two studies reported together. The first was urban and compared placentas of women of high and low socioeconomic status (SES). The women were comparable in age, parity, gestational age, absence of severe pregnancy diseases, and all the women delivered singleton males. The groups differed in income and sanitary conditions, which can be potential confounding factors of the relationship between maternal undernutrition and placental characteristics. Height, postpartum weight, triceps skinfolds and ratio of essential to nonessential serum amino acids were the nutritional indicators by which they differed. The low SES group had placentas that weighed less than those of the upper SES level but the placental chemical composition was similar except for less hydroxyproline and less fat in the low SES group. Given that women differed in other characteristics that could explain the placental attributes, a second, rural, study was conducted where supplementation with calories and proteins were given ad libitum to participating women. These women were comparable in characteristics that could affect placental weight, except for the dietary intake. Women getting more than 20,000 kcal of supplementation during pregnancy (a cut off point defined in a former study on this population as differentiating the risk of delivering a LBW infant) had heavier placentas than women getting less than 20,000 kcal. There was a statistically significant relationship between placental weight and supplementation. Unlike in the urban study, the placentas of the two rural groups had similar chemical composition in all the analysed components. Using regression methodology, the authors found a statistically significant relationship between placental and birth weight when rural and urban data were combined and SES and supplementation were controlled in the analysis.

To study the variability of BW explained by placental weight, after controlling for some indicators of maternal nutritional status (MNS) two regression models were analysed in the study (which are those proposed earlier in this paper): (1) BW = mean BW + MNS; (2) BW = mean BW + MNS + placental weight. The interest was to analyse the reduction of the proportion of the BW variance explained by the maternal nutritional status when placental weight was taken into account. The results show that there was indeed a reduction in this explanatory power (from 14 to approximately 3%). The variability of BW explained by placental weight after controlling for MNS was not reported. It is not clear to what extent the placental weight had an effect on BW (over and above nutritional status), nor if it was a better predictor of BW. A reduction in the explained variance of one indicator could have been brought about by multicollinearity among the independent variables. It will be interesting to analyse the data further to address these issues.

The study concluded that maternal undernutrition explains a significant portion of the variability of placental weight but not its analysed chemical composition (except for hydroxyproline and fat in the urban group), and that 'this effect, in turn, may be the mechanism by which maternal malnutrition causes high prevalence of low birth weight babies in these populations'.

Because this approach is good but insufficient to study the placental role in the etiology of LBW, the authors discuss the need for future research to study the effects of the possible altered placental structures that would limit fetal growth.

Rush et al. [1984] studied the effect of supplementation during gestation in a group of indigent black New York women at risk (not necessarily nutritional) of delivering LBW infants. The study randomly assigned the women to groups receiving supplements high in proteins and calories, high in calories alone or to a control group. An excess rate of prematurity among newborns whose mothers received the high protein supplementation compared to those receiving low protein supplementation or to the control was observed. There were no differences in the placentas of these groups. Women who received only extra calories had somewhat heavier babies (effect not statistically significant) and their placentas had less intervillous fibrin deposition than the other two groups of women. The authors suggest that this might mean better placental perfusion but, as was mentioned earlier, there can be up to 30% obstruction of placental surface with fibrin deposition without an impairment of its function. The functional significance of these changes is not immediately clear.

Manipulation of maternal diet and observation of placental function in vivo has been done in animals, and the results might shed some light into this research question. Bell et al. [1986] have extensively studied the role of the placenta in IUGR of sheep fetuses (although differences in placentas across species should be kept in mind when trying to make inferences to women). The authors studied heat stressed ewes early in pregnancy and subsequent fetal growth. Heat stress in the early stage of pregnancy presumably causes a redistribution of blood away from the uterus and towards the skin (for heat liberation) which impairs normal growth of the placenta. After a defined period, heat stress was removed and the animals were treated normally. Fetal growth was impaired in these fetuses as compared to control animals. The placentas of the stressed animals were smaller and had a decreased clearance of diffusable solutes. Similar work with carunculectomy in ewe placentas has also shown stunted fetal growth [Bell et al., 1986]. The proposed mechanisms are: Placental insult (heat stress or carunculectomy) → inadequate blood flow to the uterus and placenta → smaller placental size → inadequate clearance of diffusable solutes → small fetal size.

This work elegantly shows an independent effect of the placenta on sheep fetal growth. It is not clear, though, how these results apply to humans. In the ewe, the maternal tissues (caruncles) grow during pregnancy, and an early restriction could impair growth of this maternal tissue on which placenta is inserted, possibly impairing its future growth. Conversely, human maternal tissue does not grow, in fact it is eroded by the placenta. Thus, the exactly same events may not be happening in the human pregnancy. However, the work of Bell and coworkers also shows a deficient nutrient clearance in heat stressed placentas. This suggests that early restrictions might also cause deficiencies in the placental architecture impairing its function in such a way as to have an independent effect on LBW. The authors also suggest that these placentas could have an abnormal metabolism such that it directs less blood flow to the conceptus through hormonal alterations, worsening the supply line.

Present research is being conducted to study placental membranes of mothers delivering small infants [Sanchez-Griñan, 1987]. Placental diffusion capacity and its relationship with maternal malnutrition and LBW is being studied. The results of this and similar research could clarify the role of the placenta in the etiology of LBW in humans.

D. Hemodynamics during Pregnancy
1. Hormonal Control

The association between malnutrition and IUGR is complex and the linking mechanisms often have to be constructed from different studies both in humans and in animals.

Some studies have suggested an association between altered hemodynamic processes taking place during pregnancy and maternal malnutrition [Pirani et al., 1973; Ribeiro et al., 1982; Rosso and Kava, 1980; Rosso, 1981], as will be later discussed.

a) Normal Hemodynamic Changes during Pregnancy. During the course of a normal pregnancy, women experience plasma volume expansion which progressively increases starting at the end of the third month. The expansion reaches its maximum at approximately 34 to 36 weeks of pregnancy showing a terminal plateau when the mean plasma volume is about 54% larger than the nonpregnancy value, from near 2.5 to 3.5 liters [Pirani et al., 1973; Hytten and Chamberlain, 1980]. The etiology of this increase is not precisely understood. It has been suggested that in order to maintain a normal perfusion to all organs while accomodating the needs of a pregnant uterus, a physiological adaptation takes place through an increment in the plasma volume [Goodlin et al., 1983]. Thus, total uteroplacental blood flow expands to meet the metabolic and nutritional demands of a developing fetus.

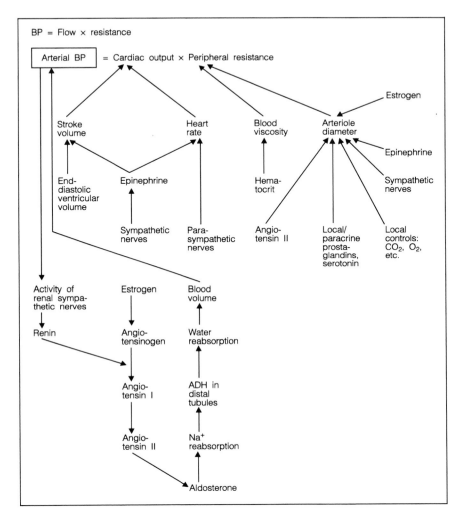

Fig. 2. Control of blood pressure (BP) and volume.

The cardiac output increases during the first half of pregnancy from 4.6 to 6 l/min. This is brought about by increased heart rate and stroke volume. Figure 2 describes the normal control of blood pressure.

The proposed mechanism for the increase of blood volume is through a vascular uteroplacental 'shunt' in combination with coordinated changes in several of the hormones secreted during pregnancy [Longo, 1983; Longo and Hardesty, 1985; Hytten and Chamberlain, 1980]. Estrogen is probably the hormone that triggers the vascular changes in pregnancy. Its production by the feto-placental unit is increased tremendously during this period (the term 'feto-placental unit' describes the cooperation of the fetus and the placenta in producing the steroid hormones during pregnancy). Cholesterol (or acetate) in maternal circulation is taken up by the placenta, and through the activation of the desmolase enzymatic system it produces the precursor pregnenolone reaching the fetal adrenal glands. There is an especially large zone in the fetal adrenal glands that synthezises dehydroepiandrosterone sulfate (DHEAS), which is hydroxylated by the fetal liver and transported back to the placenta for final aromatization and desulfation to produce estriol, secreted into the maternal circulation. Estrogens have a number of actions, one of which is to promote the synthesis of angiotensinogen by the liver (estrogen may also stimulate renin production [Longo, 1983]). This brings about elevated levels of angiotensin I and II, elevating the systemic arterial pressure. This, coupled with high levels of renin (also produced by the placenta) brings about a cascade of reactions leading to aldosterone secretion and sodium and water retention, to finally increase extracellular fluid volume.

Normally, there is a negative feedback of increased renin-increased angiotensin which, in elevated levels, decreases the production of renin. Such a mechanism is non-existent in pregnancy (it is not known precisely why) when high levels of angiotensin and renin coexist [Hytten and Chamberlain, 1980]. This phenomenon is probably related to the placental production of renin, the control of which may be different to that operating in the kidney. Changes in these hormones are related to increases of extracellular water retention.

At the same time, the placenta produces human placental lactogen (hPl) which among other functions has an erythropoietic effect, as has prolactin [Longo and Hardesty, 1985; Jepson, 1968]. Consequently there is an increase in the red cell mass of around 30% accompanying the expansion of plasma volume. So the increase in blood volume is the product of increases in plasma and in red cell mass.

High levels of estrogens and a high ratio of estrogen to progesterone elicit uterine vascular vasodilation, thereby increasing uteroplacental blood flow. This could cause a fall in the systemic blood pressure if it were not for the increased blood volume [Hytten and Chamberlain, 1980; Resnik et al., 1974] (see figure 3).

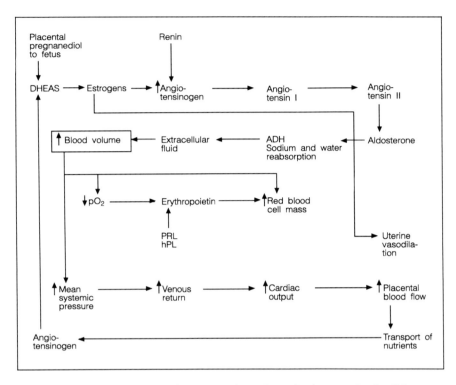

Fig. 3. Hormonal control of pregnancy hemodynamics (see text for details).

b) Observations in Undernutrition. It has been postulated that one of the biological mechanisms by which IUGR takes place is through an insufficient maternal plasma volume expansion [Rosso, 1980; 1981]. Evidence for a reduced plasma volume expansion in malnourished mothers is given by the same author [Rosso, 1978] in animal studies. Dietary restricted rats experienced either a reduced expansion or no expansion at all in their plasma volume during pregnancy.

It has also been observed that the normal increase in cardiac output does not take place in calorie and especially protein restricted pregnant rats [Rosso, 1978; Rosso and Kava, 1980]. Secondary to this impairment there is a decreased placental perfusion limiting the nutrient supply and thus the growth of the conceptus [Rosso and Kava, 1980]. Figure 4 proposed by Rosso, shows the suggested sequence of events.

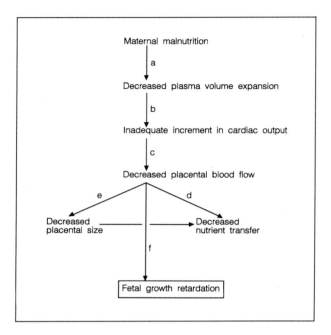

Fig. 4. Postulated chain of hemodynamic events linking maternal malnutrition and fetal growth retardation. From Rosso [1980].

In a similar animal dietary study Ahokas et al. [1983], using radioactively labelled microspheres, observed that food-restricted pregnant rats failed to increase normally their cardiac output, relative to well-fed controls. Unlike the Rosso and Kava [1980] study, the rats of this experiment showed also a decrease in the percentage of cardiac output delivered to the uteroplacental unit (link c in figure 4). This percentage increased to normal levels when the dietary restriction finished during the last week of gestation, without a significant change in total cardiac output. Increased blood flow to the uteroplacental unit was due to a marked decrease in the uteroplacental resistance of the replenished rats, which was not exhibited to the same extent by the restricted non-replenished rats. The authors suggest that the decreased vascular vasodilation in malnourished rats is probably estrogen-related. Support for this evidence comes from a human study where the urinary estrogen was reduced in undernourished Indian mothers, increasing after nutritional supplementation [Iyengar, 1968].

As Ahokas et al. [1983] note, the fact that the uteroplacental blood flow increased in the absence of an increased cardiac output suggests that the latter might not be a necessary prerequisite to increase placental perfusion.

A follow-up investigation by Ahokas et al. [1986] found that administration of an alpha-adrenergic receptor blocker to food restricted rats, resulted in significant increases in uterine perfusion.The same treatment given to rats fed ad libitum did not alter placental perfusion. This suggest that dietary restriction is a stressful condition which elevates circulating levels of vasoconstrictor catecholamines. Interference with the actions of catecholamines without a change in the diet leads to normal hemodynamics in underfed rats, indicating that hemodynamic changes associated with malnutrition might not be directly caused by low nutritional intake, but to stress-related endocrine events.

The former Ahokas et al. [1983] study shows that fetal weight was not significantly different in the products of replenished rats exhibiting increased uteroplacental blood flow when compared to those fed ad libitum throughout gestation, as opposed to dietary restricted rats, which had significantly lighter pups.

Some other animal studies also give support to the proposed relationship between placental blood flow and placental/fetal weight (d, e, and f links in figure 4). In chronically catheterized guinea pigs [Peeters et al., 1982] placental and fetal size were linearly correlated with placental blood flow as measured with radioactively labelled microspheres (links d and f). The extent to which these results apply to humans is not known. However, the study of the processes that take place in mammalian pregnancy during malnutrition brings light to the mechanisms of human IUGR.

Some human studies suggest an impairment in normal gravid-related hemodynamic processes during maternal undernutrition. In a study of 11,082 patients from the collaborative perinatal project of the US National Institue of Neurological and Communicative Disorders and Stroke, Naeye [1981] reports a positive statistically significant relationship between prepregnant body mass and maximum diastolic blood pressure after mid gestation (MDBP), and also between weight gain during pregnancy and the same blood pressure measurement. Furthermore, in women without edema or proteinuria a statistically significant relationship was found between MDBP and BW, up to a diastolic pressure of 90 mm Hg, above which BW levelled off. The higher the prepregnant ponderal index and the larger the maternal weight gain (under 8 kg), the larger BW in relation to increasing MDBP below 90 mm Hg. Head circumference at birth was related to mean diastolic blood pressure above 90 mm Hg, except in very low maternal weight gain (< 2 kg).

The fact that blood pressure is positively related to uteroplacental blood flow, which in term is associated with BW was studied in humans by Grunberger et al. [1979] and Lunnel et al. [1979]. Both groups measured uteroplacental blood flow by injecting into the maternal circulation indium-113, a short-lived radio tracer which binds with maternal transferrin and does not cross the placenta. Images were taken by gamma scintillation camera, and blood flow indices were derived from these measurements. Lunnel et al. [1979] observed in a small sample size (n = 19) a strong statistically significant relationship between IUGR and uteroplacental blood flow indices.

Grunberger et al. [1979] prospectively studied 70 low blood-pressured pregnant women (BP < 100/65, selection of cut-off point not explained) and treated those who had severe symptoms of low blood pressure with deoxycorticosterone trimethylacetate (DOCTA) until blood pressure rose to normal levels. Thirteen women did not experience an increased blood pressure despite treatment. DOCTA increases sodium retention (weak mineralocorticoid effect) and stimulates epinephrine secretion, which is a mild vasoconstrictor but that increases considerably the cardiac output. Before the administration of DOCTA, 80–90% of the women had subnormal uteroplacental blood flow. After the administration of DOCTA to 40 women, their blood flow indices improved markedly, and only 17% of these women still had inadequate values. The mean BW observed in the neonates of the untreated mothers was 2,860 g as opposed to 3,308 g of those of the treated women. There is not much information about the maternal characteristics but no mother was reported to be malnourished or otherwise ill. The fact that uteroplacental blood flow and birth weight increased after increments in blood pressure suggests that there might be a causal relationship between these variables.

In another study, Ribeiro et al. [1982] were interested in the nutritional component of the hemodynamic adaptations during pregnancy. The authors studied the relationship between postpartum systolic blood pressure (SBP) and pregnacy starvation in the women exposed to the Dutch famine of 1944–1946. The objective of the study was to analyse whether a relationship existed between low blood pressure at delivery and low prevalence of preeclampsia. Using regression methodology, the authors observed a statistically significant relationship between mean caloric intake

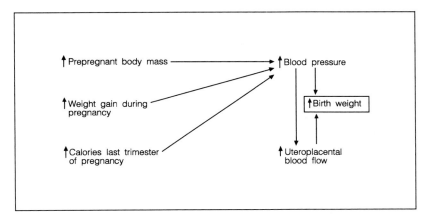

Fig. 5. Malnutrition, hemodynamics, and birth weith: some observed associations in pregnant women (see text for details). These relationships do not imply causality. Blood pressure relationships observed in hypotensive or normotensive women, and not in hypertensive women.

during the three months before delivery (which reached a maximum of 2,200 kcal) and SBP at delivery in two famine-striken cities. In one of the two cities studied, the relationship found was linear and positive, and in the second city the relationship was apparently curvilinear showing a statistically not significant threshold of the hypertensive effect at about 1,600 kcal/day, above which SBP decreased. However, the amount of variance of SBP explained by caloric intake was not more than 7% in either group. Ten-day-postpartum weight (not concurrently taken with SBP) failed to explain a significant proportion of the variance of SBP at delivery, after controlling for the average caloric intake during the three months before delivery. This suggests that the observed relationship between caloric intake and SBP was not through maternal net weight. The lack of association between weight and blood pressure is at odds with the widely accepted concept that these variables are related in nonpregnant subjects [Earnst, 1984]. Ribeiro and coworkers [1982] explain that this may be due to physiological postpartum changes that might have taken place in the interval between measurements of blood pressure and weight. Some of the discussed associations are depicted in figure 5.

With regards to plasma volume expansion, there are no human studies linking maternal malnutrition during pregnancy and a reduced expansion.

Nevertheless, it has been repeatedly observed [Pirani et al., 1973; Hytten and Paintin, 1963; Hytten and Chamberlain, 1980; Longo, 1983; Longo and Hardesty, 1985] that a relationship exists between maternal blood volume expansion during pregnancy and birth weight.

c) Conclusions. Blood volume expansion has been observed to be altered in acutely or chronically malnourished pregnant animals, and it is accompanied by IUGR.

Although blood volume expansion during pregnancy has been studied in humans, and has ben related to birth weight, inadequate increase of blood volume has not been studied in human maternal undernutriton. Nevertheless, alterations of other hemodynamic processes such as low blood pressure have been observed in undernourished pregnant women. It remains to be studied if hormonal production is altered in undernourished women, and if these alterations are the causal factor of the altered hemodynamics observed in these women. In figure 3, where does malnutrition intervene so as to disrupt the process? How can we overlap figures 3 and 5?

It is not clear whether and how a better nutritional status of pregnant women is associated with changes in blood pressure, blood volume expansion, or hormonal production. It remains to be observed if undernourished mothers experience an inadequate blood volume expansion (as has been observed in rats), and if birth weight is reduced in these women's infants. A final comment on this subject is that an absolute decrease in plasma nutrient concentration cannot be ruled out as the only causal factor of IUGR without altering placental or hormonal pregnancy physiology.

III. Maternal Undernutrition and Lactation Performance

The functional consequences of maternal malnutrition go beyond the pregnancy interval into lactation; and this issue is the focus of the following section.

Physiological and behavioral mechanisms determine milk output through a balance of complex factors. These factors are related to the size of the infant (and this with age, sex and BW), the breastfeeding patterns and the nutritional status of the mother. Other variables such as maternal age and parity also exert an influence on the amount of milk produced, but their impact is relatively small.

A. Physiology of Lactation

Milk output directly depends on the action of the hormones prolactin and oxytocin. Nipple stimulation during suckling brings about an increase in circulating prolactin levels through a neural arc from the nipple to the pituitary [Lawrence, 1980]. Prolactin is the hormone responsible for milk synthesis in the mature alveoli cells of the mammary gland for the next three to four hours of its synthesis [Lawrence, 1980]. Once milk is synthesized in the mammary gland, it needs to be removed. Indeed, the opportune removal of milk from the alveolar ducts is indispensable for the continuation of lactation. Oxytocin, produced by the hypothalamus and stored in the neurohypophysis, is released to the bloodstream at each feeding session through the stimulation of the mechanoreceptors located on the nipple. In women, conditioned reflexes such as the sight or sound of the baby, and feelings of confidence trigger also oxytocin release. The action of this hormone makes milk, already stored in the mammary gland, available to the infant through its contractile effect on the ducts.

Anxiety can impair this neural pathway in such a way that adequate amounts of oxytocin are not released, and the ducts are not completely emptied [Jelliffe and Jelliffe, 1979; Hadley, 1984].

Thus, physiological, psychological and behavioral factors are involved in milk production through the frequency and duration of suckling occasions as well as through psychosomatic factors of the mother.

B. Lactation Performance in Undernourished Women

Numerous observational studies have reported that women with undernutrition produce smaller amounts of milk than their well-nourished counterparts [Hanafy et al., 1972; Van Steenbergen et al., 1983; Naing et al., 1980; Jelliffe and Jelliffe, 1979]. Roughly, undernourished women produce 500–700 ml/day, but the range is very wide.

1. Biological Mechanisms

As discussed above, biological as well as behavioral factors explain milk output, so that the relationship between nutritional status and milk production should be analysed taking into account potential confounding factors.

Maternal undernutrition, both before and during pregnancy, is independently associated with LBW as discussed in former sections (being LBW a functional consequence of maternal undernutrition). On the other hand, the size of the infant is positively associated to the amount of milk the mother produces [Roberts et al., 1982; Jelliffe and Jelliffe, 1979]. Thus, LBW might be the biological mechanism by which undernourished mothers produce small volumes of milk, most probably through less vigorous sucking or less frequent demand from a LBW baby. In this context, the mother's potential to produce milk does not play a role.

Another way in which undernutrition may limit milk production is through a reduced availability of nutrients for milk production at the alveolar cell level. Although this makes biological sense, it is most difficult to prove in humans. To prove that the physiological capacity to produce milk is limited by undernutrition requires the demonstration that milk production increases after an improvement in nutritional status of undernourished lactating women. Observational studies give strong suggestion that this might happen, but intervention studies must be done to answer this question.

There are certain conditions in the research design that must be met in order to address this issue adequately. The women in such studies should be producing small amounts of milk and be either undernourished or have a low dietary energy intake explaining at least part of the low milk output. In other words, only the women with potential to respond to a food supplementation can respond.

A small number of supplementation studies during lactation have been conducted over the past 30 years. These have been done in metabolic wards and on the home of small groups of women. Only a few have been done at the community level.

The studies conducted in metabolic wards have been done in Nigeria [Edozien et al., 1976] and India [Gopalan, 1958; Karmarkar et al., 1963]. None of the three studies give data about the nutritional status of the women, although all reported that the subjects were malnourished. Women received protein or calories plus protein in supplements offered for various periods of time during lactation. Milk volumes were measured by the test-weighting technique from 1 to 3 consecutive days (in the Nigerian study, milk volume was determined by test-weighting plus hand expression of the milk remaining in the ducts after each suckling episode). All the studies show an impact of supplementation on milk volume (differences ranging from 10 to 240 ml/day). Improvements in milk composition related to supplementation were observed in the studies in India; the Nigerian study does not report milk composition. These positive studies should be taken as suggestive, rather than a proof of an impact of supplementation on milk production. This is so because the increase in dietary intake was not the only variable that was changed. Restful conditions prevail in metabolic wards, especially under the circumstances described in the studies. Patterns of breast feeding were also altered, and all these changes may account to some degree for the differences observed between the groups or after supplementation.

Community controlled experimental trials are important methodological improvements in the field of nutritional epidemiology. Through these, stronger evidence to construct the causal link between human nutrition and functional outcomes is constructed. Furthermore, if the studies are randomized and double-blind, causality can be infered. A small number of community intervention trials have been conducted with mixed results. None of the studies have been randomized or double-blind. Results from a supplementation study conducted in Mexico in a group of rural women [Chávez and Martínez, 1980] show that an increased intake of calories and protein by poor women with short stature was followed by an increased volume of milk production of decreased concentration. Caloric content of milk samples was not determined, but the authors concluded that it remained unchanged after the supplementation. In Gambia, Prentice et al. [1980] carried out a supplementation trial in lactating women living in substistence farming and with exceedingly low caloric intakes, i.e., 1,200–1,750 kcal/day as measured by 24-hours dietary recall. Supplementing their diets with up to 1,100 kcal/day had no effect on milk production. These results have caused great controversy in the field of nutrition increasing our interest to direct future attention to more methodologically stringent supplementation trials, and to studies on changes in efficiency in the use of energy under stressful conditions. As the authors have argued, one important reason that could explain the lack of effect might be that the women were not malnourished. Their anthropometry was normal, and their work capacity (as measured by the amount of physical activity performed) was also normal. This means that although the reported recall intakes were very low in calories, the women were not truly needy. Therefore, this issue remains unresolved, namely if truly needy women increase their milk production after supplementation. This question has important policy implications because if undernourished lactating women do respond to food supplementation by increasing milk production, more milk will be available to the infant and the weaning period can be delayed. This would give the infant's immune response more time to develop and to improve its defense ability at a time when morbidity and mortality are very high, and mostly due to diarrhoeal diseases [Mata, 1978].

IV. Corollary

This document reviews the functional consequences of maternal undernutrition and possible underlying biological mechanisms. Consequences of maternal malnutrition of vast public health importance include the delivery of LBW infants and an insufficient milk production. The consequences for the child of these conditions are increased neonatal and postneonatal mortality, growth faltering, increased morbidity and impaired neurological development [Mata, 1978; McCormick, 1985; Villar et al., 1984].

The functional consequences for the mother herself include decreased work capacity, a constant general feeling of tiredness and probaby increased morbidity. This review has also pointed out that these conditions are preventable. Improving the accessibility to food commodities and to primary health care are public health interventions that could have a great impact in the survival, health and performance of an important segment of the world's population.

We hope that this document offers useful aggregated scientific information for nutrition policy makers, and their advisors who feel the need to deepen their understanding of the maternal malnutrition problems.

References

Ahokas, R.; Anderson, G.; Lipshitz, J.: Cardiac output and uteroplacental blood flow in diet-restricted and diet-replenished rats. Am. J. Obstet. Gynec. *146:* 6–13 (1983).

Ahokas, R.; Reynolds, S.; Anderson, G.; Lipshitz, J.: Catecholamine-mediated reduction in uteroplacental blood flow in the diet-restricted term-pregnant rat. J. Nutr. *116:* 412–418 (1986).

Anderson, J.D.; Blender, I.N.; McClemont, S.; Sinclair, J.C.: Determinants of size at birth in a Canadian population. Am. J. Obstet. Gynec. *150:* 236–244 (1984).

Ashworth, A.; Feachem, R.G.: Interventions for the control of diarrhoeal diseases among young children: Prevention of low birth weight. Bull. Wld. Hlth Org. *83:* 165–184 (1985).

Bell, A.W.; Bawman, D.E.; Currie, W.B.: Regulation of nutrient partitioning and metabolism during pre- and postnatal growth. Paper presented at the ASAS Biennial Symposium on Current Concepts of Animal Growth, Manhattan, N.Y., July 1986.

Black, R.E.; Brown, K.H.; Becker, S.: Malnutrition is a determinant factor in diarrhoeal duration, but not incidence, among young children in a longitudinal study in rural Bangladesh. Am. J. clin. Nutr. *39:* 87–94 (1984).

Bonds, D.; Gabs, S.G.; Kumar, S.; Taylor, T.: Fetal weight/placental weight ratio and perinatal outcome. Am. J. Obstet. Gynec. *149:* 195–200 (1984).

Brown, K.H.; Akhtar, A.N.; Robertson, A.D.; Ahmed, M.G.: Lactational capacity of marginally nourished mothers: relationship between maternal nutritional status and quantity and proximal composition of milk. Pediatrics, Springfield 78: 909–919 (1986).

Chandra, R.K.: Immunodeficiency in undernutrition and overnutrition. Nutr. Rev. 39: 225–231 (1981).

Chase, H.C.: Infant mortality and weight at birth: 1960 United States Cohort. Am. J. publ. Hlth 59: 1618–1628 (1969).

Chávez, A.; Martinez, C.: The effect of maternal supplementation on infant development. Arch. Latinoam. Nutr. 29: suppl. 1, pp. 143–153 (1979).

Chávez, A.; Martínez, C.: Effects of maternal undernutrition and dietary supplementation on milk production; in Aebi and Whitehead, Maternal nutrition during pregnancy and lactation, pp. 274–284 (Huber, Bern 1980).

Chávez, A.; Martínez, C.; Bourges, H.; Coronado, M.; López, M.; Basta, S.: Child nutrition problems during lactation in poor rural areas. Proc. IXth National Congress Nutr., México 1972, vol. 2, pp. 90–105 (Karger, Basel 1975a).

Chávez, A.; Martínez, C.; Yaschine, T.: Nutrition, behavioral development, and mother-child interaction in young rural children. Fed. Proc. 34: 1574–1582 (1975b).

Chen, L.C.; Chowdhury, A.; Huffman, S.L.: Anthropometric assessment of energy-protein malnutrition and subsequent risk of mortality among preschool-aged children. Am. J. clin. Nutr. 33: 1836–1845 (1980).

Cravioto, J.; DeLicardie, E.R.: Mental performance in school-age children. Am. J. Dis. Child. 120: 404–410 (1970).

Cravioto, J.: DeLicardie, E.R.: Mother-infant relationship prior to the development of clinically severe malnutrition in the child. Proc. West. Hemisph. Nutr. Congr. 4Th. (Miami 1975).

Cunningham-Rundles, A.: Effects of nutritional status on immunological function. Am. J. clin. Nutr. 35: 1202 (1982).

Delgado, H.; Martorell, R.; Brineman, E.; Klein, R.: Nutrition and length of gestation. Nutr. Res. 2: 117–126 (1982).

Dickin, K.: Placental characteristics, infant birth weight and nutritional status among highland and lowland Bolivian women; thesis Cornell University, Ithaca (1986).

Dougherty, G.R.S.; Jones, A.D.: The determinants of birth weight. Am. J. Obstet. Gynec. 144: 190–200 (1982).

Earnst, M.D.; Levy, R.I.: Diet and cardiovascular disease; in Olsen, Present knowledge in nutrition, 5th ed., pp. 724–739 (The Nutrition Foundation Inc., Washington 1984).

Edozien, J.; Khan, M.A.R.; Waslien, C.L.: Human protein deficiency: results of a Nigerian village study. J. Nutr. 106: 312–328 (1976).

Edwards, L.E.; Alton, I.R.; Barrada, M.I.; Hakanson, E.Y.: Pregnancy in the underweight women: course, outcome and growth patterns of the infant. Am. J. Obstet. Gynec. 135: 297–302 (1979).

Ferguson, A.C.: Prolonged impairment of cellular immunity in children with intrauterine growth retardation. J. Pediat. 93: 52–56 (1978).

Fox, H.: The correlation between placental structure and transfer function; in Chamberlain and Wilkison, Placental transfer, pp. 15–30 (Pitman Medical, New York 1979).

Goodlin, R.C.; Gobry, C.A.; Anderson, J.C.; Woods, R.E.; Quaife, M.: Clinical signs of normal plasma volume expansion during pregnancy. Am. J. Obstet. Gynec. *145:* 1001–1009 (1983).

Gopalan, G.: Studies on lactation on poor Indian communities. J. trop. Pediatr. *4:* 87–97 (1958).

Gruenwald, P.: The placenta and its maternal supply line (University Park Press, Baltimore 1975).

Grunberger, W.; Leodolter, F.; Parschald, O.: Maternal hypotension: fetal outcome in treated and untreated cases. Gynecol. obstet. Invest. *10:* 32–38 (1979).

Haas, J.; Balcazar, H.; Caulfield, L.: Variation in early neonatal mortality for different types of fetal growth retardation. Am. J. phys. Anthrop. *73:* 467–473 (1987).

Habicht, J.P.; Yarbrough, C.; Lechtig, A.; Klein, R.E.: Relationships of birth weight, maternal nutrition and infant mortality. Nutr. Rep. int. *7:* 533–546 (1973).

Hadley, M.E.: Endocrinology (Prentice-Hall, Englewood Cliffs 1984).

Hanafy, M.M.; Morsey, M.R.A.; Seddik, Y.; Habib, Y.A.; Lozy, M.: Maternal nutrition and lactation performance. Environ. Child Health. *18:* 187–191 (1972).

Hemminki, E.; Starfield, B.: Routine administration of iron and vitamins during pregnancy: review of controlled clinical trials. Br. J. Obstet. Gynec. *85:* 404–410 (1978a).

Hemminki, E.; Starfield, B.: Prevention of low birth weight and pre-term birth. Literature review and suggestions for research policy. Health and Society *56:* 339–346 (1978b).

Hytten, F.; Chamberlain, G.: Clinical physiology in obstetrics (Blackwell, London 1980).

Hytten, F.; Paintin, D.B.: Increase in plasma volume during pregnancy. J. Obstet. Gynaec. Br. Commonw. *70:* 402–407 (1963).

Institute of Medicine. Preventing low birth weight (NAS, Washington 1985).

Iyengar, L.: Urinary estrogen excretion in undernourished Indian pregnant women: effect of dietary supplement on urinary estrogens and birth weights of infants. Am. J. Obstet. Gynec. *102:* 834–838 (1968).

Iyengar, L.; Rajalakshmi, K.: Effect of folic acid supplementation on birth weights of infants. Am. J. Obstet. Gynec. *122:* 332–336 (1975).

Jepson, G.H.: Endocrine control of maternal and fetal erythropoiesis. Can. med. Ass. J. *98:* 844–848 (1968).

Jelliffe, D.B.; Jelliffe, E.F.P.: Human milk in the modern world, pp. 59–83 (Oxford Medical Publications, Oxford 1979).

Kaiser, I.H.: Fertilization and the physiology and development of fetus and placenta. In Danforth, D.N., Obstetrics and Gynecology; 3rd ed., pp. 240–270 (Harper & Row, New York 1977).

Kaminski, M.; Goujard, J-Rumeau-Rouquette C.: Prediction of low birth weight and prematurity by multiple regression analysis with maternal characteristics known since the beginning of the pregnancy. Int. J. Epidemiol. *2:* 195–204 (1973).

Karmarkar, M.J.; Rajalakshmi, R.; Ramakrishanan, C.V.: Studies on human lactation. I. Effect of dietary protein and fat on protein, fat and essential amino acid content in breast milk. Acta paediat. scand. *52:* 473–480 (1963).

Kramer, M.S.: Determinants of intrauterine growth and gestational duration. A methodological assessment and synthesis. Manuscript from the Departments of Pediatrics and of Epidemiology and Biostatistics, McGill University Faculty of Medicine,

Montreal, Canada, and the Maternal and Child Health Unit and Nutrition Unit, World Health Organization, Geneva, Switzerland (1985).

Kramer, M.S.: Intrauterine growth and gestational duration determinants. Pediatrics *80:* 502–511 (1987).

Laga, E.M.; Driscoll, S.G.; Munro, H.N.: Comparison of placentas from two socioeconomic groups. 1: Morphometry. Pediatrics *49:* 24–32 (1972).

Lawrence, R.A.: Breast-feeding. A guide for the medical profession, pp. 28–43 (Mosby, St. Louis, 1980).

Lechtig, A.; Delgado, H.; Yarbrough, C.; Habicht, J.P.; Martorell, R.; Klein, R.E.: A simple assessment of the risk of low birth weight to select women for nutritional intervention. Am. J. Obstet. Gynec. *125:* 25–34 (1976).

Lechtig, A.; Habicht, J.P.; Delgado, H.; Klein, R.E.; Yarbrough, C.; Martorell, R.: Effect of food supplementation during pregnancy on birth weight. Pediatrics *56:* 508–520 (1975a).

Lechtig, A.; Klein, R.: Pre-natal nutrition and birth weight: Is there a causal association? in Dobbing, Maternal nutrition in pregnancy – eating for two? pp. 131–156 (Academy Press Univ. Manchester (1980a).

Lechtig, A.; Klein, R.; Guía para interpretar la ganancia de peso durante el embarazo como indicador de riesgo de bajo peso al nacer. Bol. Of. Sanit. Panam. *89:* 489–495 (1980b).

Lechtig, A.; Yarbrough, C.; Delgado, H.; Habicht, J.P.; Martorell, R.; Klein, R.: Influence of maternal nutrition on birth weight. Am. J. clin. Nutr. *28:* 1223–1233 (1975b).

Lechtig, A.; Yarbrough, C.; Delgado, H.; Martorell, R.; Klein, R.; Behar, M.: Effect of moderate maternal malnutrition on the placenta. Am. J. Obstet. Gynec. *123:* 191–201 (1975c).

Longo, L.D.: Maternal blood volume and cardiac output during pregnancy. A hypothesis of endocrinologic control. Am. J. Physiol. *245:* R720–R729 (1983).

Longo, L.D.; Hardesty, J.S.: Maternal blood volume: measurement, hypothesis of control, and clinical considerations. Rev. Perinat. Med. *5:* 35–59 (1985).

Lunnell, N.O.; Sarby, B.; Lewander, R.; Nylund, R.: Comparison of uteroplacental blood flow in normal and in uterine growth-retarded pregnancy. Gynecol. obstet. Invest. *10:* 106–118 (1979).

Martorell, R.; González-Cossío, T.: Maternal nutrition and birth weight. Yb of Phys. Anthrop. *30:* 195–220 (1987).

Mata, L.J.: The children of Santa María Cauqué: A prospective field study of health and growth (MIT Press, Cambridge 1978).

McCormick, M.C.: The contribution of low birth weight to infant mortality and childhood morbidity. New Engl. J. Med. *312:* 82–90 (1985).

McDonald, E.C.; Pollit, E.; Mueller, W.; Hsueh, A.M.; Sherwin, R.: The Bacon chow study: Maternal nutritional supplementation and birth weight of offspring. Am. J. clin. Nutr. *34:* 2133–2144 (1981).

Metcoff, J.; Costiloe, P.; Crosby, W.; Sandstead, H.H.; McClain, P.; Bodwell, C.E.; Bentle, L.; Seshachalam, D.: Maternal malnutrition and fetal outcome. Am. J. clin. Nutr. *34:* 708–721 (1981).

Moore, K.: The developing human: clinically oriented embryology; 2nd ed., pp. 85–107 (Saunders, Philadelphia 1977).

Mora, J.O.; de Paredes, B.; Wagner, M.; de Navarro, L.; Suescun, J.; Christiansen, N.;

Herrera, M.G.: Nutritional supplementation and the outcome of pregnancy. 1. Birth weight. Am. J. clin. Nutr. *32:* 455–462 (1979).

Morris, F.: Placental factors conditioning fetal nutrition and growth. Am. J. clin. Nutr. *34:* 760–768 (1981).

MSPAS, INCAP. Informe Final. Encuesta nacional simplificada de salud y nutrición materno infantil (Guatemala 1986).

Munro, H.N.: Placenta in relation to nutrition. Fed. Proc. *39:* 236–238 (1980).

Murthy, L.S.; Agarwal, K.N.; Khanna, S.: Placental morphometric and morphological alterations in maternal undernutrition. Am. J. Obstet. Gynec. *124:* 641–646 (1967).

Naeye, R.L.: Nutritional/nonnutritional interactions that affect the outcome of pregnancy. Am. J. clin. Nutr. *34:* 727–731 (1981a).

Naeye, R.: Maternal nutrition and pregnancy outcome; In Dobbing, Maternal nutrition in pregnancy – eating for two?, pp. 89–102 (Academic Press, London 1981b).

Naeye, M.D.: Maternal blood pressure and fetal growth. Am. J. Obstet. Gynec. *141:* 780–787 (1981).

National Academy of Sciences (NAS): Recommended Dietary Allowances; 9th revised ed. Committee on Dietary Allowances, Food and Nutrition Board, Division of Biological Sciences, Assembly of Life Sciences. National Research Council (Washington 1980).

Naing, Khin-Maung; Tin-Tin-Oo; Kywe-Thein; Nwe-New-Hlaing.: Study of lactation performance of Burmese mothers. Am. J. clin. Nutr. *33:* 2665–2668 (1980).

Peeters, L.L.H.; Sparks, J.W.; Grutters, G.; Girard, J.; Battaglia, F.C.: Uteroplacental blood flow during pregnancy in chronically catheterized guinea pigs. Pediat. Res. *16:* 716–720 (1982).

Pirani, B.B.K.; Campbel, D.M.; MacGillivray, I.: Plasma volume in normal first pregnancy. J. Obstet. Gynaec. Br. Commonw. *80:* 884–887 (1973).

Prentice, A.M.; Whitehead, R.G.; Roberts, S.B.; Paul, A.A.; Watkinson, M.; Prentice, A.; Watkinson, A.A.: Dietary supplementation of Gambian nursing mothers and lactational performance. Lancet *ii:* 886–888 (1980).

Prentice, A.M.; Whitehead, R.G.; Watkinson, M.; Lamb, W.H.: Prenatal dietary supplementation of African women and birth weight. Lancet *i:* 489–491 (1983).

Resnik, R.; Kielman, A.; Battaglia, C.; Makowski, E.L.; Meschia, G.: The stimulation of uterine blood flow by various estrogens. Endocrinology *94:* 1192–1196 (1974).

Ribeiro, M.D.; Stein, Z.; Susser, M.; Cohen, P.; Neugut, R.; Prenatal starvation and maternal blood pressure near delivery. Am. J. clin. Nutr. *35:* 535–541 (1982).

Rivera, J.: Cellular immune response in moderate malnutrition; thesis Cornell University, Ithaca (1984).

Roberts, S.B.; Paul, A.A.; Cole, T.J.; Whitehead, R.G.: Seasonal changes in activity, birth weight and lactational performance in rural Gambian women. Trans. R. Soc. trop. Med. Hyg. *76:* 668–678 (1982).

Rosso, P.: Maternal nutritional status and plasma volume expansion in the pregnant rat (Abstract). Fed. Proc. *37:* 491 (1978).

Rosso, P.: Placental growth, development and function in relation to maternal nutrition. Fed Proc. *39:* 250–254 (1980).

Rosso, P.; Kava, R.: Effect of food restriction on cardiac output and blood flow to the uterus and placenta in the pregnant rat. J. Nutr. *110:* 2350–2354 (1980).

Rosso, P.: Nutrition and maternal fetal exchange. Am. J. clin. Nutr. *34:* 744–755 (1981).

Rush, D.; Kristal, A.; Navarro, C.; Chauhan, P.; Blank, W.; Naeye, R.; Susser, M.W.: The effects of dietary supplementation during pregnancy on placental morphology, pathology, and histomorphometry. Am. J. clin. Nutr. *39:* 863–871 (1984).

Rush, D.; Stein, B.; Susser, M.A.: A randomized controlled trial of prenatal nutritional supplementation in New York City. Pediatrics *65:* 683–697 (1980).

Sánchez-Griñán, M.I.: Maternal malnutrition and placental diffusion performance. A stereological study. Research proposal (Cornell University, Ithaca 1987).

Tanner, J.M.: Fetus into man: physical growth from conception to maturity, pp. 40–43 (Harvard University Press, Cambridge 1978).

Van Steenbergen, W.M.; Kusin, J.A.; Dewith, C.; Lacko, E.; Jansen, A.A.J.: Lactation performance of mothers of contrasting nutritional status in rural Kenya. Acta paediat. scand. *72:* 805–810 (1983).

Villar, J.; Belizán, J.M.: The timing factor in the pathophysiology of the intrauterine growth retarded syndrome. Obstetl gynec. Surv. *37:* 499–506 (1982).

Villar, J.; Smeriglio, V.; Martorell, R.; Brown, C.H.; Klein, R.E.: Heterogeneous growth and mental development of intrauterine growth retarded infants during the first 3 years of life. Pediatrics *74:* 783–791 (1984).

Waterlow, J.C.; Thompson, A.M.: Observations on the adequacy of breastfeeding. Lancet *ii:* 238–242 (1979).

Weller, R.H.; Bouvier, L.F.: Population; Demography and Policy. St. Martin's Press (New York 1981).

Winikoff, B.; Debrovner, C.: Anthropometric determinants of birth weight. Obstet Gynecol. *58:* 678–684 (1981).

World Health Organization, Div. Fam. Health: The prevalence of nutritional anemia in women in developing countries (WHO (FHE/79.3), Geneva 1979).

Teresa González-Cossío, MSc, Division of Nutrition and Health, Institute of Nutrition of Central America and Panamá, Guatemala City (Guatemala)

Subject Index